Elements of Two-Stroke Engine Development

SP-988

GLOBAL MOBILITY DATABASE

*All SAE papers, standards, and selected
books are abstracted and indexed in the
Global Mobility Database.*

Published by:
Society of Automotive Engineers, Inc.
400 Commonwealth Drive
Warrendale, PA 15096-0001
USA
Phone: (412) 776-4841
Fax: (412) 776-5760
September 1993

4-97

#29216294

ISBN 1-56091-424-6
SAE/SP-93/988
Library of Congress Catalog Card Number: 93-85055
Copyright 1993 Society of Automotive Engineers, Inc.

PREFACE

Elements of Two-Stroke Engine Development (SP-988) is a collection of seven technical papers presented during two sessions held at the 1993 SAE International Off-Highway & Powerplant Congress & Exposition in Milwaukee, Wisconsin, September 13-16, 1993. SP-988 covers a diverse range of issues related to the development of two-stroke-cycle engines and should be of interest to those working in the field.

The first paper, one of five papers presented by the group at the Queen's University of Belfast (QUB), examined the emissions of a simple 0.05-L carbureted single cylinder two-stroke engine. A method to reduce the loss of short-circuited fuel during scavenging is described, whereby air is introduced into the top of the transfer passages during crankcase induction. Several different port arrangements were studied. Significant reduction in fuel consumption and hydrocarbon emissions were achieved at wide-open-throttle operation, but performance deteriorated under part-load operation.

Ongoing work to use a Stirling engine to recover energy from the exhaust gas of a two-stroke engine was reported in the second paper. A carbureted two-stroke engine incorporated an exhaust-port design that enabled the exhaust gas flow to be split into two flows, one with a high concentration of unburned and the other with a high concentration of burned gas. The burned-gas flow was treated with a conventional three-way catalyst. The flow having a high concentration of unburned gas was temporarily stored in a tank, burned in an afterburner, and some energy was recovered with a Stirling engine. In this way the overall efficiency of the system was increased.

The third paper examined preignition with high-olefin fuels, which lead to in-cylinder carbon deposits, in a 0.15-L air-cooled scooter engine. An oil which gave no preignition with a reference fuel was used. The tests showed that an increase in gasoline olefin content gave an increased propensity to preignition, particularly for levels above 20%. When the olefin content was high, no improvement in preignition behavior was found when a fuel additive designed to reduce carbon deposits was added.

Preliminary experiments with direct air-assist gasoline injection to a cross-scavenged single-cylinder 0.5-L engine were reported in the fourth paper. The engine incorporated a deflector-type piston designed at QUB. Parameters varied included exhaust-port timing, injection timing, and spark timing. Wide-open-throttle tests at 3000 r/min showed good brake mean effective pressures, but operation at a 1600-r/min part-load condition were less satisfactory.

In the fifth paper, the effects of flaw maldistribution on the performance of catalytic converters was examined in an experimental flow apparatus. The focus was converters for small capacity carbureted two-stroke engines. Inlet cone designs that might be suitable for motorcycle applications were evaluated. Two different flow-distribution models were interfaced with a mathematical model containing kinetics expressions describing catalyst behavior. A flow-distribution ratio was defined that could be correlated to catalyst efficiency with the catalyst model.

Operation of a 0.43-L single-cylinder two-stroke engine on biogas (60-70% methane) was described in the sixth paper. The research engine was to be a potential retrofit unit for existing multi cylinder four-stroke stationary power generators. The paper cites several reasons why a two-stroke engine is well suited to biogas operation. Biogas was supplied directly to the cylinder with a low pressure gas injector. The initial test results presented for the engine illustrate the excellent potential of the concept.

In the final paper, an extension of a non-isentropic unsteady gas-dynamics model based on the method of characteristics, published previously by Professor Blair of QUB, was presented. The extension addresses new theory for gradual and sudden area changes, and the results of applications to several diffuser sections. The computed results were judged to be good by the authors, and experimental work is underway at QUB to verify the results.

Another recent SAE publication addressing two-stroke-cycle engines, *Advanced Two-Stroke Engines* (SP-942), is also recommended.

Edward G. Groff
GM NAO R&D Center

Benjamin L. Sheaffer
Mercury Marine

TABLE OF CONTENTS

Reduction of Fuel Consumption and Emissions for a Small Capacity Two-Stroke Cycle Engine

S.J. Magee, R. Douglas, and G. P. Blair
The Queen's University of Belfast

J-P. Cressard
Peugeot MTC

ABSTRACT

The emissions produced from a simple carburetted crankcase scavenged two-stroke cycle engine primarily arise due to losses of fresh charge from the exhaust port during the scavenging process. These losses lead to inferior fuel consumption and a negative impact on the environment. Pressure on exhaust emissions and fuel consumption has reduced the number of applications of the two-stroke cycle engine over the years, however the attributes of simplicity, high power density and potential low manufacturing costs have ensured its continuing use for mopeds and motorcycles, small outboard engines and small utility engines. Even these last bastions of the simple two-stroke engine are being challenged by the four stroke alternative as emissions legislation becomes tighter and is newly formulated for many categories of engines. A simple solution is described which reduces short circuit and scavenge losses in a cost effective way. Air is introduced into the top of the transfer passages during the crankcase induction phase, this air is then used to initiate the scavenging process. A preliminary experimental evaluation has shown that brake specific fuel consumption can be reduced by about 10% and brake specific hydrocarbon emissions by around 40% at wide open throttle operation.

INTRODUCTION

At QUB continual research is being under taken in the field of engine modelling using unsteady gas dynamics and high levels of confidence can be placed on the simulation code (1,2)*. If a time history of the purity of the gas at the exhaust port is plotted against crank angle a characteristic s-curve is seen as shown in Fig. 1. The sudden increase in the fuel concentration at the exhaust port is due to the arrival of significant amounts of fresh charge. If it is assumed that the charge which arrives at the exhaust port is that which has been resident in the cylinder the longest then,

if a scavenging buffer is used, a reduction in scavenging losses should result. This system has been employed on loop-scavenged engines(3,4,5,6,7) quite recently and more historically on classical cross-scavenged engines (8,9). It is now suggested that a cross-scavenged engine is the most suitable for the application of stratified scavenging or 'Air-Head' scavenging using secondary air intakes at the top of the transfer passages controlled by reed valves.

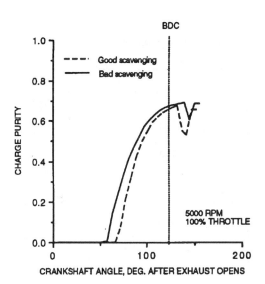

Fig. 1 Purity at Exhaust Port

The schematic shown in Fig. 2 illustrates the principle of operation of an air-head stratified scavenging system applied to a QUB cross-scavenged engine. During the induction stroke, Fig. 2a, a mixture of air and fuel is

* Numbers in parenthesis designate references at the end of this paper.

induced into the crankcase while simultaneously air only is induced into the top of the transfer passages through the auxiliary air inlet. The air displaces any air/fuel mixture in the transfer passages remaining from the previous scavenging process. When the transfer ports open the scavenging process is initiated with pure air. This is then followed by a rich air/fuel mixture from the crankcase. The assumption that significant losses of fresh charge take place from the first portion of charge to enter the cylinder suggests that an air-head system should reduce fuel losses to the exhaust and thus improve fuel consumption and emissions.

At QUB a research program was initiated to evaluate under firing conditions a 50 cc QUB deflector piston engine compared to a loop scavenged design, then to determine the improvements which could be achieved using a stratified scavenging system.

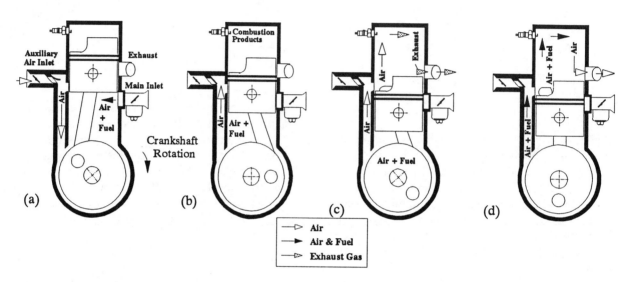

Fig. 2 Air-Head Operation

	QUB 50 cc CROSS	50 cc LOOP
Type	Two-stroke	Two-stroke
Induction System	Reed valve	Reed valve
Cooling System	Liquid	Fan assisted air
Bore	40mm	40mm
Stroke	39.1mm	39.1mm
Con-Rod Length	80mm	80mm
Trapped Comp. Ratio	6.7:1	6.6
Exhaust Port Opens	105 atdc	98.2 atdc
Transfer Port Opens	120 atdc	123.5 atdc
Deflection Ratio	1.11	
Exhaust System	Semi Tuned Silencer	
Fuel System	Diaphragm Carburettor	
Lubrication System	Crankshaft Driven Pump metered to inlet passage	

Table 1   Engine Specifications.

ENGINE SPECIFICATION AND EXPERIMENTAL APPARATUS USED.

In order to make a comparison between the various engine configurations it was considered necessary that both engines should be designed to have the same power characteristic. The main specifications for both engines are given in table 1.

The engines were directly coupled to a Borghi and Saveri eddy current dynamometer, air consumption was measured at the carburettor and at the auxiliary air intake throttle body (when in use) by means of the pressure drop across a BS1042 orifice plate (10) at the entrance to a large baffled surge tank. Fuel flow rate was measured by a Max series 210 positive displacement flow meter. All performance calculations were calculated to SAE Test Code J1349 (11) and corrected to STP using an ISO (12) correction factor. A Cussons p8333 exhaust gas analyzer was used to analyse the emissions. This equipment employed a Servomex 155a paramagnetic oxygen detector and NDIR detection for HC, CO and CO_2. The exhaust gas was sampled from the mid-section of the exhaust system before the first silencer box. Brake specific HC emission results were calculated as detailed in ref (13).

SCAVENGE TESTING OF ENGINE CYLINDERS.

One prerequisite thought necessary for the application of stratified scavenging was that the secondary air inlets and reed valves would not be intrusive to the flow in the transfer passages and thus not deteriorate the scavenging. A new transfer passage configuration was thus designed as shown in Fig. 4. This layout was significantly different from anything tested at QUB and before firing tests were carried out an investigation of the scavenging efficiency was carried out using the QUB single cycle scavenging rig (14).

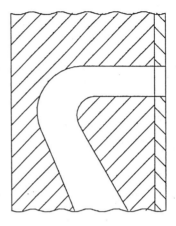

Fig. 4 QUB Air-Head Transfer Passage

SCAVENGE TESTS PERFORMED. A 50 cc QUB deflector piston cylinder was designed with "conventional" QUB type scavenge passages as shown in Fig. 3. These consist of two parallel vertical passages which intersect with horizontal slots machined into the cylinder barrel and through the liner to form the port. A cover plate is then inserted to block the outer part of the machined passages. It is contoured to aid the passage of the scavenge flow into the cylinder. The obvious means of applying a stratified scavenging system is to omit this cover plate and affix a reed valve and throttle body to the outside of the cylinder barrel (15,16) as shown in Fig. 5.

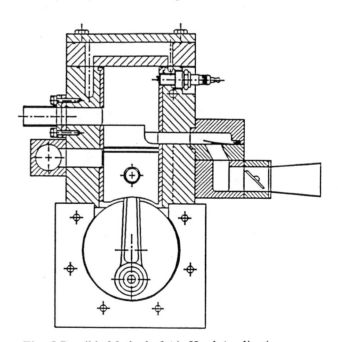

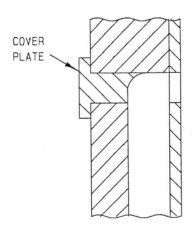

COVER PLATE

Fig. 3 Conventional QUB Deflector Transfer Passage

Fig. 5 Possible Method of Air-Head Application

To investigate the effect on scavenging, a flat cover plate was used at the proposed position of the reed valve. The new transfer passage was considerably different from preceding designs but had two main advantages. The passages meant that in theory if there was no mixing, the maximum effect could be gained from the scavenge buffer by delaying the entrance of fresh charge to the cylinder for as long as possible. The reed valves are also situated in such a position as to minimise their effect on the scavenge flow, however concern was expressed as to the amount of secondary air that it would be possible for them to flow.

TEST NOMENCLATURE

Design 1   Two port QUB cross scavenged 170 cc design with curved piston, coverplate as shown in Fig. 3

Design 2   Two port QUB cross scavenged 50 cc design with curved piston, coverplate as shown in Fig. 3

Design 3   Two port QUB cross scavenged 50 cc Air-Head design A, flat coverplate used.

Design 4   Two port QUB cross scavenged 50 cc Air-Head design B with curved transfer passages as shown in Fig. 4. Port arrangement as in Fig. 6

Design 5   Two port QUB cross scavenged 50 cc Air-Head design C with curved transfer passages as shown in Fig. 4. Port arrangement as in Fig. 7

Design 6   Five port loop scavenged 50 cc design.

The results of these tests are presented in Fig. 8.

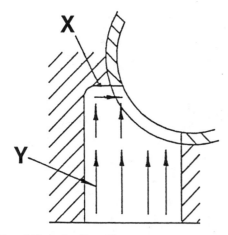

Fig. 6 Transfer Port Entrance

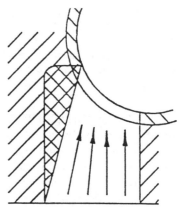

Fig. 7 Modified Transfer Port Entrance

ENGINE TEST PROCEDURE.

The duty cycles of small internal combustion engines are many and varied but, excluding idling operation, the most important condition for emissions and fuel consumption is the wide open throttle point. The first tests were performed on the 50 cc loop scavenged engine at wide open throttle and 3000-9000 rpm. The same test was performed on the cross scavenged engine to determine a reference before the QUB Air Head scavenging system was implemented.
The auxiliary air ports were then opened to their full extent and the corresponding tests were performed in a stratified scavenging mode. A part throttle test was then performed at the maximum torque speed of 6500rpm. During this test the carburettor throttle and the air throttle were closed simultaneously to try to ensure a constant difference in air flow. This procedure was felt necessary in order to try to simulate the equivalent condition of throttles connected by an adjustable linkage. For all tests the engines were equipped with a diaphragm carburettor which was adjusted at each test point to achieve a trapped air fuel ratio of around 12 to 1. Results from these tests are shown in Figs. 9,10 and 11.

RESULTS AND DISCUSSION

SCAVENGE TESTS. The first test performed on the standard deflector was quite disappointing as it did not compare favourably with previous tests performed on a 170 cc QUB cross scavenged cylinder as seen in Fig. 8a. The 170 cc engine had a bore-stroke ratio of 1.11 compared to 1.02 for the 50 cc engine. Normally it would be expected that a more square cross scavenged engine would scavenge better than an over square engine (18). However, traditionally smaller cylinders do not scavenge as well as larger ones; this entire subject area has been the topic of a comprehensive research program at this university and will be reported at SAE at a later date. Thus it would appear that the smaller capacity of the cylinder had a greater bearing on its scavenging than its bore-stroke ratio. Fig. 8b shows the significance of the contoured cover plate, design

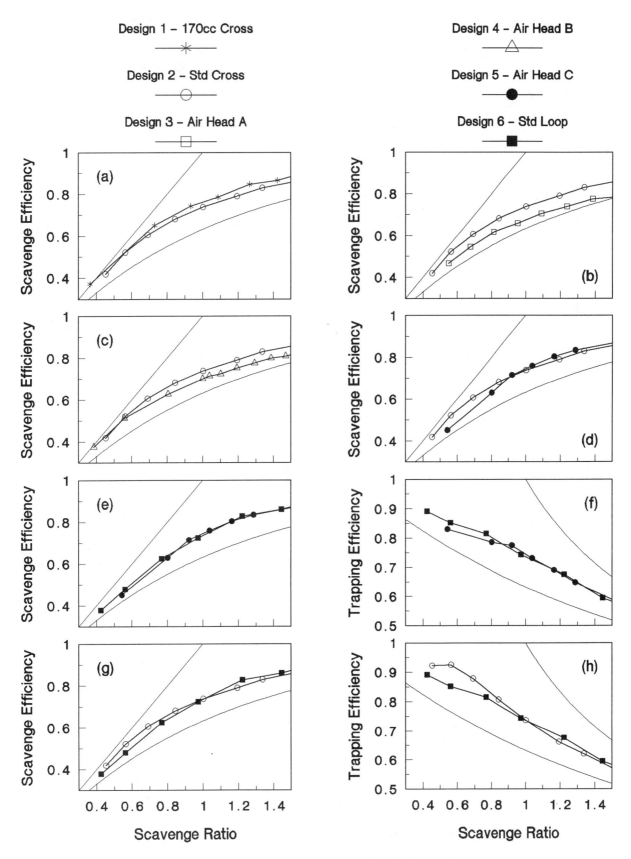

Figure 8   Scavenge Test Results

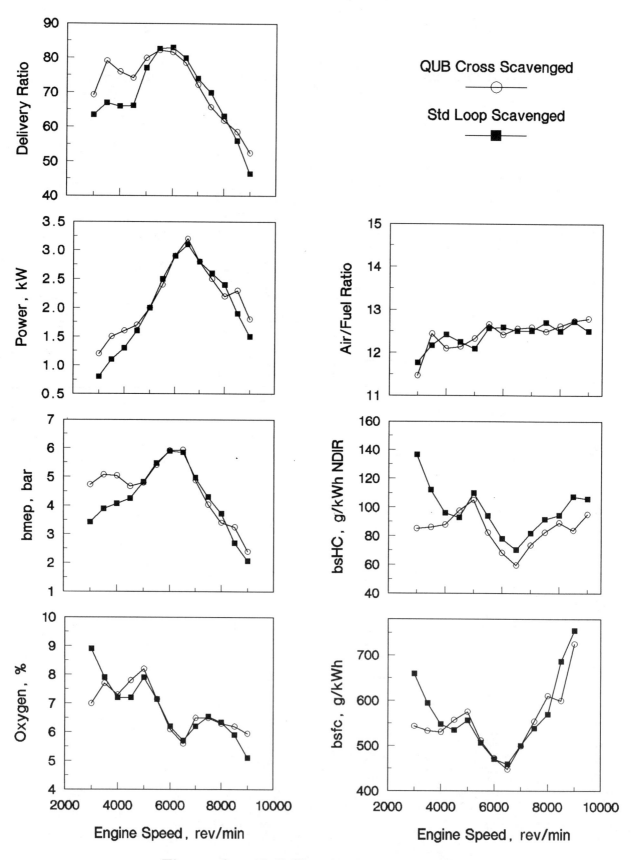

Figure 9   Full Throttle Test Results

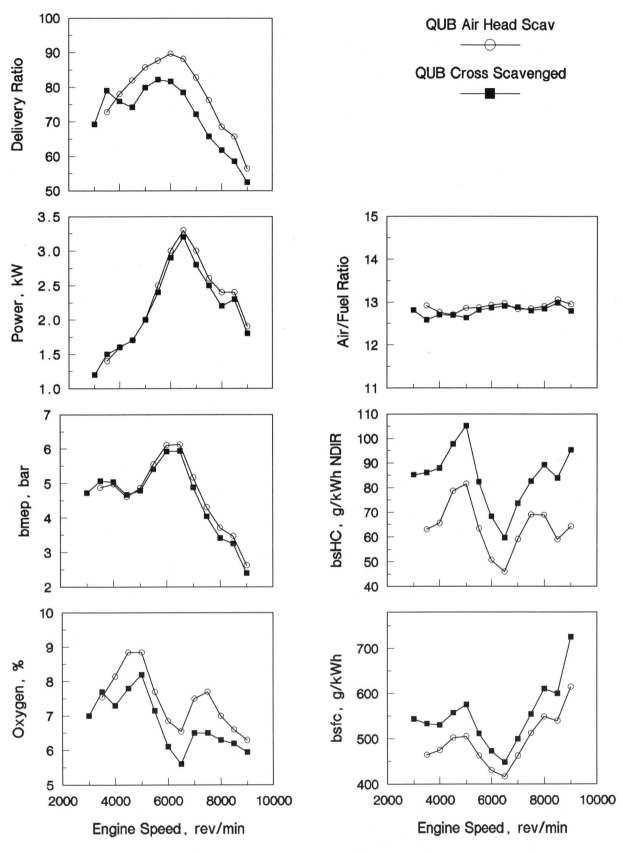

Figure 10   Full Throttle Tests

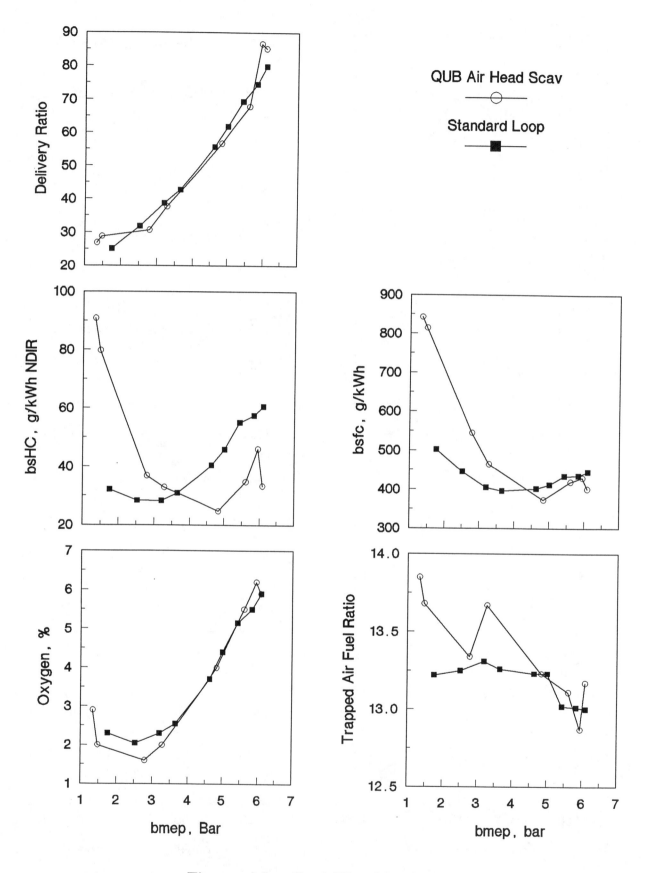

Figure 11   Part Throttle Tests

2 having very inferior scavenging to design 1 as would be expected due to the possibility of excessive turbulence at the port and a recirculating flow regime in the transfer passage at the right angle bend. This result shows that if the cover plate was removed and the secondary air introduced through the opening the scavenging would be seriously deteriorated.

The shape of the graph of purity at the exhaust port is the most important factor in determining the scavenge system most suited to application of a scavenge buffer system. At this point of the program, it was felt that it would be most prudent to start with as good a scavenging characteristic as possible and thus this configuration was not pursued.

The scavenging performance of the new design of transfer passages was inferior to the standard vertical passages shown by Fig. 8c. Consideration was made to the passage of air from the transfer duct to the cylinder, and it was noticed that the air labelled Y would impinge on the wall labelled X shown in Fig. 6. This means that the flow is unlikely to be well directed and any sort of radial cohesive flow impossible, thus a modification was effected to promote a smoother path for the gas as shown in Fig. 7. The change was dramatic, as shown in Fig. 8d. The lines now cross at a scavenge ratio of about 1 however, the scavenging efficiency of the Air-Head design C is still noticeably inferior to the conventional QUB cross below this value.

It is worthwhile to consider the significance of scavenge ratio as determined from the single cycle gas scavenging rig. This is explained by Blair (2), its use in engine modelling and the relevance to the firing engine performance. The most important considerations are, the experiment is performed under isothermal isovolumetric conditions, and the scavenge ratio and scavenge efficiency are calculated by volume. The corresponding measurements made on a running engine are by mass, yet initially the scavenging flow expands by heating during the open cycle. A scavenge ratio by mass in reality often expands to double the value by volume, thus scavenge ratios up to 1.4 as determined from the scavenge rig have an important effect on the engine performance.

The comparison of design 5 and design 6, Figs. 8e and 8f show that, except for the two lowest values of scavenge ratio, the QUB Air-Head design C is at least as good as the 50 cc loop scavenged design. This was a very good result as previous work (17) had shown the 50 cc loop scavenged design to have very good scavenging. Figs. 8g and 8h make the comparison between the vertical passages on the QUB cross scavenged engine and the loop scavenged design. It highlights the change in scavenging characteristic when the flow is allowed to enter the cylinder of a QUB deflector piston engine with zero elevation angle. The scavenging performance more closely resembles a loop scavenged engine. It has poor scavenging below a scavenge ratio of 0.8 but very good scavenging above this value.

ENGINE TEST RESULTS. The performance of both engines is remarkably similar, the main differences being at engine speeds below 4500rpm. Below this speed the QUB cross scavenged engine has a significantly higher delivery ratio, approximately 15% higher. This combined with much better trapping of fresh charge at the lowest speeds, (as can be seen from the oxygen concentration in Fig. 9), gives a substantial bmep and torque advantage. As a result of the better scavenging, the brake specific fuel consumption and hydrocarbon emissions are not deteriorated, they are much superior with 10% better fuel consumption and 40% brake specific hydrocarbon emission improvement at 3000rpm. Over the rest of the speed range, the performance of the engines is close enough for any discrepancy to be due to experimental error. The shape of the power curve is typical for the type of partially tuned exhaust pipe which was used. These results compliment those obtained from the single shot scavenge rig, which predicted very similar scavenging performance.

RESULTS WITH AIR-HEAD SYSTEM. In Fig. 10 the performance of the QUB cross scavenged engine with and without the secondary air inlet throttle fully open is presented. The delivery ratio through the engine is shown in Fig. 1. In this design the majority of the air enters the engine through the carburettor. At higher speed the proportion of air entering through the air-head changes significantly. The total delivery ratio characteristic of the engine is consistently higher with the air throttle open due to the reduced restriction to the air flow. The scavenge ratio of the engine is deteriorated due to the higher delivery ratio, but slightly more power is produced at all speeds above 5500rpm as a result of a higher charging efficiency. As the engine is now operating in a stratified scavenging mode, the overall air-fuel ratio calculated from air flow and fuel flow is no longer indicative of the air fuel ratio trapped in the cylinder. The trapped air fuel ratio was calculated from emissions measurements of oxygen and hydrocarbons present which represent the unburnt air and fuel exiting the engine. The "trapping efficiency" of air and fuel can be assessed and hence the burning zone AFR. A complete explanation of the calculation procedure is given in (12). From Fig. 10 it can be seen that the trapped air fuel ratio of both engines was kept very close throughout the speed range. The improvement in fuel consumption of 9% at all speeds translates to a reduction in brake specific hydrocarbon emissions of 40%. This is considered to be a very significant improvement in engine performance with very little development work. The proportion of air delivered through the air throttle is not optimised and was the value obtained when the air throttle was fully opened.

PART THROTTLE TESTS. The delivery ratio requirement of the two engines for equivalent power is practically the same, this combined with similar oxygen values, suggests a comparable charge utilisation. The QUB cross engine exhibits particularly bad performance characteristics at bmep values less than 3.5 bar. When the total delivery ratio is so low, it is suspected that not enough fuel is consistently reaching the cylinder promoting cycle to cycle variation in air fuel ratio. Although the engine was run at each speed for a significant amount of time to preclude transient measurement, it is suggested that fuel could build up in the crankcase exacerbating this effect. The emissions equipment would average the emissions, giving a false reading of air fuel ratio. The argument thus arises for the complete closure of the air throttle below a certain value of delivery ratio. Above a the 3.5 bar point the stratified charged engine again exhibits superior hydrocarbon emissions and slightly better fuel consumption.

The part throttle tests highlight a potential difficulty in the application of this system. At part throttle operation and light loads the proportion of air through the secondary air inlet is particularly important, the excessive induction of secondary air being detrimental to the engine performance. The detailed design of the intake system, carburation and flow partitioning are the key design areas. Work must be done to simulate the operation of the system in order to try to predict the required air quantities. This, coupled with further steady state and dynamic testing, has the potential to produce small utility engines capable of passing future emissions legislation.

CONCLUSIONS.

1. A 50 cc loop scavenged and a QUB cross-scavanged engine were constructed. The engine performance was very similar above 4500 rpm, with the QUB cross-scavenged engine exhibiting marginally better characteristics at the lowest engine speeds.

2. The application of an air-head type stratified scavenging system is more easily applied to a cross-scavenged engine.

3. The application of an air-head type stratified scavenging system to a 50 cc QUB cross-scavenged engine resulted in a considerable improvement in fuel consumption and hydrocarbon emission. The bsfc was improved by approximately 9%, throughout the speed range at wide open throttle and maximum secondary air flow, accompanied by a 40% reduction in bsHC, relative to the standard QUB cross scavenged engine.

4. The improvement in fuel consumption and brake specific hydrocarbon emissions of a 50 cc engine have been shown to be similar to that found by other researchers working on larger capacity engine at equal operating conditions (15, 16).

5. At light load, part throttle settings the air-head system considerably deteriorated bsfc and bsHC. Careful consideration will have to be paid to throttle synchronisation to obtain the optimum air flow through each inlet.

6. The air-head system is an attractive solution for emissions reduction where the predominant engine operating condition is at wide open throttle.

ACKNOWLEDGEMENTS.

The authors would like to thank The Queens University of Belfast for the provision of laboratory and workshop facilities. The assistance of Mr R McCullough and workshop technicians are also acknowledged without whom this work would have been impossible. Technical assistance from Peugeot MTC (France) is also gratefully acknowledged.

REFERENCES.

1.   G.P. Blair, "An Alternative Method for the Prediction of Unsteady Gas Flow through the Internal Combustion Engine", Society of Automotive Engineers International Off-Highway & Powerplant Congress, Milwaukee, Wisconsin, September 9-12, 1991, SAE Paper No. 911850.

2.   G.P. Blair, "Correlation of an Alternative Method for the Prediction of Engine Performance Characteristics with Measured Data", Society of Automotive Engineers International Congress and Exposition, Detroit Michigan, March 1-5, 1993, SAE Paper No.930501.

3.   K. Blume, "Zweitakt-Gemischspulung mit Spulvorlage", Motortechnische Zeitschrift, No.74, Vol. 12, pp. 475-478, 1972.

4.   F.Wolf, "Gemischgespulte Zweitakt-Brennkrafmaschine", German Patent No. 2650834, July, 1977.

5.   T. Iwai, "Two-Cycle Internal Combustion Engines", Yamaha Hatsudoki Kabushiki Kaisha, Iwata, Japan, U.S. Patent No. 4075985, February, 1978.

6.   A. Ramesh, B. Nagalingam, K.V. Gopalakrishnan, "Improvement In The Performance Of A Two-Stroke Spark-Ignition Engine Through Extra Reed Valves Fitted At The Transfer Ducts", SAE Paper No. 859374, 1985.

7.   M. Saxena, H.B. Mathur, S. Radzimirski, "A Stratified Charging Two-Stroke Engine for Reduction of Scavenged-Through Losses", Society of Automotive Engineers Small Engine Technology Conferance, Milwaukee, Wisconsin, September 11-13, 1989, SAE Paper No. 891805.

8.   W. Stephenson, "Explosive Engine", U.S Patent No. 1012288, December, 1911.

9.   F.W. Lanchester, R.H. Pearsall, "An Investigation of Certain Aspects of the Two-Stroke Engine for Automobile Vehicles", The Automobile Engineer, February, 1922.

10. BS1042, "Fluid Flow in Closed Conduits", British Standards Institution, 1981.

11. SAE J1349, Engine Power Test Code, Spark Ignition and Diesel, June 1985.

12. ISO3046, Reciprocating Internal Combustion Engines: Performance-Parts 1, 2 and 3, International Standards Organisation, 1981.

13. R. Douglas, "AFR and Emission Calculations for Two-Stroke Cycle Engines", SAE Paper No. 901599, September 1990.

14. M.E.G Sweeney, R.G. Kenny, G.B.G. Swann, G.P. Blair, "Single Cycle Gas Testing Method for Two-Stroke Engine Scavenging", SAE Paper No. 850178, February 1985.

15. R.J. Kee, G.P. Blair, C.E. Carson, R.G. Kenny, "Exhaust Emissions of a Stratified Charge Two-Stroke Engine", 5th Graz Two-Wheeler Symposium, April 1993.

16. C.E. Carson, R.J. Kee, R.G. Kenny, G.P. Blair, "The Reduction of Emissions From Two-Stroke Engines", Second International Seminar on Worldwide Emission Standards and How To Meet Them, IMechE Headquarters, London, May 25-26, 1993.

17. R.G. Kenny, "Peugeot 50 cc Scavenging", Report No.1875, Dept of Mechanical and Industrial Engineering, The Queens University of Belfast, March 1990.

18. R.J. Kee, G.P. Blair, R. Douglas, "Comparison of Performance Characteristics of Loop and Cross Scavenged Two-Stroke Engines", International Off-Highway & Powerplant Congress and Exposition, Milwaukee, Wisconsin, September 10-13, 1990, SAE Paper No. 901666.

932394

Effective Energy Utilization and Emission Reduction of Exhaust Gas in a Two-Stroke Cycle Engine — Part II

Kazuo Sato, Masamitsu Nakano, and Haruo Ukawa
Shibaura Institute of Technology

ABSTRACT

This paper deals with a research project concerning an effective utilization of exhaust gas heat. Exhaust gas from a exhaust gas-separate type two-stroke cycle engine containing a high concentration of unburned gas was temporarily stored in a floating-bell type tank as an form of heat energy conservation, while in the previous report [1]* exhaust heat was recovered with continuous operation. A Stirling engine with a hot-water supply system was then used to oxidize or burn again the exhaust gas in a catalyzer and an after-burner unit in order to recover the unspent heat energy from the exhaust gas. A three-way catalyzer was employed to remove pollutants both from the combustion gas in this process and the high-concentration burned gas from the two-stroke cycle engine.

The results of the research in the present paper are intended as a follow-up of the previous report [1] to clarify a method for the more effective use of exhaust gas heat. Similarly to the results presented in the previous paper the present follow-up report establishes a method for the nearly complete removal of the pollutants CO and HC from emission gas at a fuel lean range of air-fuel ratio. The method significantly reduces the NO content of the gas at the same air-fuel ratio. It should also be noted that utilizing the Zeldovich reaction mechanism increases the ability to verify NO formation characteristics in the emission gas for measured results. The previous paper had reported that the bimolecular reaction mechanism also served to study the formation characteristics of NO in the process.

INTRODUCTION

A series of our studies on exhaust gas from a two-stroke engine have been conducted by dividing the exhaust gas into two components. One component contained a high-concentration of unburned gas while the other contained a high-concentration of burned gas. The division of the components was used to examine how heat energy in each of these components could be effectively reused and to determine how the pollutant emission of such gases into the environment could be reduced.

To demonstrate that exhaust gas could be separated into two components, a separately installed four-stroke cycle engine was operated on a high-concentration unburned gas mixture which was contained in the exhaust gas from a two-stroke cycle engine [2]. This study was subsequently supplemented by a study which investigated the operating conditions, power output performance, pollutant gas reducing effect, and other features of the four-stroke cycle engine when operated with the useable part of gas separated from the exhaust of the two-stroke cycle engine [3], [4]. It was then undertaken to use the separated high-concentration burned gas, i.e. the high-temperature exhaust gas, as a heat source.

* Numbers is parentheses designate references at the end of the paper.

A separately installed Stirling engine was operated with this gas in order to determine how useful the recycling of exhaust gas could be [5]. The Stirling engine was also operated using heat from the combustion of a high-concentration unburned gas in an after-burner. The high-temperature gas was treated by a catalyzer with further clarifying the most effective method for recovering the useable heat energy in exhaust gas. This treatment also served to prove the effectiveness of the process for almost completely eliminating the pollutants HC and CO from the emission gas [6].

It was furthermore undertaken to simulate how such pollutant gases as HC, CO and NO are produced in the two-stroke cycle engine. An after-burner was used for the high-concentration unburned gas while a three-way catalyzer was employed for the treatment of high-concentration burned gas. The results revealed that computer simulation provides a reliable means of examining how effectively pollutants can be removed from exhaust gases. The results also revealed that with the proposed method the catalyzer together with an after-burner unit is able to almost completely remove CO and HC from emission gas at a fuel lean range of air-fuel ratio and significantly reduce the NO content of the gas [7].

The present paper proceeds to report the results of a study in which exhaust with a high-concentration of unburned gas is reburned. The exhaust having a high-concentration of unburned gas from a two-stroke cycle engine was temporarily stored in a floating-bell type tank, which was used as a temporal heat reservoir without using the continuous heat recovery method such as the previous report. Subsequently a Stirling engine with a hot-water supply system was used to recover heat-energy from the exhaust gas by burning it again in an after-burner unit.

A three-way catalyzer was employed to clarify the burned gas used in this process of emission gases, i.e. air pollutants. It was also used for removing the same pollutants from the high-concentration burned gas emitted from the exhaust port of the two-stroke cycle engine.

The use of the bimolecular reaction mechanism for the estimation of NO has already been expounded in our previous paper. The present follow-up study will point out that the Zeldovich reaction mechanism may also be used to estimate the content of NO in exhaust gas. The Zeldovich reaction mechanism enabled us to discern more clearly than in the previous paper the more effective method for utilizing exhaust gas heat. Equally as indicated by the results of the previous paper the follow-up study also formulates a method for almost completely removing the pollutants CO and HC from emission gas whithin a lean range of air-fuel ratio and for eliminating the NO content within the same range of the air-fuel ratio. The Zeldovich reaction mechanism also provided us with a more effective method for verifying NO estimation for measured value.

COMBUSTION

COMPOSITION OF COMBUSTION GAS IN A TWO-STROKE CYCLE ENGINE — The present paper calculates combustion gas products according to the following assumptions:

(1) Except for NO, all products of combustion are in a chemical equilibrium at 1,000 K or more, while they are in a state of frozen equilibrium at less than 1,000 K.

(2) The calculation accounts for the following eleven compositions: O_2, N_2, CO_2, H_2O, CO, H_2, OH, NO, O, H, and N.

(3) The fuel used for the engine is a hydrocarbon in the form of C_mH_n and the air used for combustion is a mixture of N_2 and O_2.

(4) All chemicals involved are perfect gases.

The excess air ratio is given as λ ($= \gamma / \gamma_0$) the mole fraction ratio of N_2 to O_2 in air is given as c ($= 79.01/20.99$), and the mole number of combustion gas compositions ( ) is given as $n_{(\ )}$. The total mole number is given as n_{er}, and the concentrations of CO and O_2 existing even in frozen equilibrium are given as $(CO)_c$ and $(O_2)_c$. Under these stipulations the combustion equation can be formulated as follows:

$$C_mH_n + \frac{\gamma}{\gamma_0}(m+\frac{n}{4})O_2 + \frac{79.01}{20.99}\frac{\gamma}{\gamma_0}(m+\frac{n}{4})N_2$$
$$= n_{O2}O_2 + n_{N2}N_2 + n_{CO2}CO_2 + n_{H2O}H_2O + n_{CO}CO$$
$$+ n_{H2}H_2 + n_{NO}NO + n_{OH}OH + n_OO + n_HH + n_NN$$
$$+ (CO)_c n_{er}CO + (O_2)_c n_{er}O_2$$

$$\cdots\cdots(1)$$

Accordingly the following equations can be formulated from the mass balance of four atoms immediately before and after the occurrence of the reaction:

$$N_C = n_{CO_2} + n_{CO} + (CO)_c n_{er}$$
$$N_H = 2n_{H_2O} + 2n_{H_2} + n_{OH} + n_H$$
$$N_O = 2n_{CO_2} + n_{H_2O} + 2n_{O_2} + n_{OH} + n_{NO} + n_{CO}$$
$$+ (CO)_c n_{er} + 2(O_2)_c n_{er}$$
$$N_N = 2n_{N_2} + n_{NO} + n_N \quad \cdots\cdots\cdots\cdots (2)$$

Where $N_C = m$, $N_H = n$, $N_O = 2(\gamma/\gamma_0)(m + n/4)$, and $N_N = 2(79.01/20.99)(\gamma/\gamma_0)(m + n/4)$; under the state of the frozen equilibrium, $n_O = 0$, $n_{OH} = 0$, $n_{O_2} = 0$, $n_H = 0$, and $n_N = 0$, for a fuel rich mixture range, while $n_O = 0$, $n_{OH} = 0$, $n_H = 0$, $n_{CO} = 0$, and $n_N = 0$ for a fuel lean mixture range.

The calculation also assumes that the following dissociation reactions are to be found in combustion products:

$$CO_2 \Leftrightarrow CO + \frac{1}{2}O_2$$
$$H_2O \Leftrightarrow H_2 + \frac{1}{2}O_2$$
$$OH \Leftrightarrow \frac{1}{2}H_2 + \frac{1}{2}O_2$$
$$O \Leftrightarrow \frac{1}{2}O_2$$
$$H \Leftrightarrow \frac{1}{2}H_2 \quad \cdots\cdots\cdots\cdots (3)$$

In the above reactions, the equilibrium constant $k_{P()}$ for CO_2, for instance, can be written as follows:

$$k_{PCO_2} = \frac{n_{CO}n_{O_2}^{1/2}}{n_{CO_2}}\left(\frac{R'T}{V}\right)^{1/2} \quad \cdots (4)$$

Bimolecular Reaction —— Examination of NO formation is based on the supposed non-equilibrium bimolecular reaction mechanism that consists of two molecules, N_2 and O_2. Given the following from reaction kinetics theory:

$$N_2 + O_2 \overset{k_f}{\underset{k_b}{\rightleftharpoons}} 2NO \quad \cdots\cdots\cdots\cdots (5)$$

It follows:

$$\frac{dn_{no}}{dt} = \frac{1}{V}[k_f \cdot n_{N_2}\{n_{O_2} + (O_2)_c \cdot n_{er}\}$$
$$- k_b \cdot n_{NO}^2] \quad \cdots\cdots\cdots\cdots (6)$$

When Kaufmann's factors [8] are used for the reaction the rate constants k_f and k_b are given as follows:

$$k_f = 5.48 \times 10^{10} \cdot \exp(-107060/RT)$$
$$k_b = 2.6 \times 10^9 \cdot \exp(-63800/RT)$$
$$[\ell/mol \cdot sec]$$

Zeldovich Reaction —— For comparing NO estimation values with the reaction mechanism supposed in previous report, the Zeldovich reaction, another mechanism of reaction for NO formation, is also used as an assumption for this calculation. From reaction kinetics theory, it follows:

$$N_2 + O \overset{k_{1f}}{\underset{k_{1b}}{\rightleftharpoons}} NO + N$$
$$O_2 + N \overset{k_{2f}}{\underset{k_{2b}}{\rightleftharpoons}} NO + O$$

$$\frac{dn_{NO}}{dt} = \frac{1}{V}\left(A + \frac{C \cdot D}{B}\right) \quad \cdots\cdots\cdots\cdots (7)$$

$$A = k_{1f} \cdot n_{N_2} \cdot n_O - k_{2b} \cdot n_{NO} \cdot n_O$$
$$B = k_{1b} \cdot n_{NO} + k_{2f}\{n_{O_2} + (O_2) \cdot n_{er}\}$$
$$C = -k_{1b} \cdot n_{NO} + k_{2b}\{n_{O_2} + (O_2) \cdot n_{er}\}$$
$$D = k_{1b} \cdot n_{N_2} \cdot n_O + k_{2b} \cdot n_{NO} \cdot n_O \quad \cdots\cdots (8)$$

In the above reactions the rate constants k_{1f}, k_{1b}, k_{2f}, and k_{2b} are given Newhall's values [9]:

$$k_{1f} = 7 \times 10^{10} \cdot (-75500/RT)$$
$$k_{1b} = 1.55 \times 10^{10}$$
$$k_{2f} = 13.2 \times 10^6 \cdot T \cdot \exp(-7080/RT)$$
$$k_{2b} = 3.2 \times 10^6 \cdot T \cdot \exp(-39100/RT)$$
$$[\ell/mol \cdot sec]$$

The gas constants R' and R are given as $R' = 0.08205$ [atm·ℓ/mol·K] and $R = 1.987$ [cal/mol·K]. The calculation also uses $(O_2)_c = 0.004$ and $(CO)_c = 0.004$ based on our experimental result [10]. The volume V in equations (4), (6), (7) is given as a function of the crank angle while the equilibrium constant k_{PCO_2} in equation (4) is given as a function of the temperature T. Under these conditions the mole number of each gas composition can be determined by solving simultaneously equations (1) through (8).
From the pressure P'_{za}, temperature T'_{za} and volume V'_{za} at the start of the compression stroke and the scavenging efficiency η_s, the mole number of each gas composition can be determined from the following process.

First, the total mole number n'_{za} in the cylinder can be written as:

$$n_{za} = \frac{P'_{za} \, V'_{za}}{R' \, T_{za}} \qquad \cdots\cdots\cdots (9)$$

If the mole numbers before and after the combustion reaction in equation (1) are given as n_{fa} and n_{er} respectively, the ratio R_n of the total mole number of gases after the combustion reaction n'_{za} to gases before the combustion reaction in the cylinder is:

$$R_n = \frac{n'_{za}}{\eta_s n_{fa} + (1 - \eta_s) n_{er}} \qquad \cdots\cdots (10)$$

On the basis of equation (10), therefore, the composition of burned gas, O_2 for instance, can be written as:

$$n_{O2} = R_n \{ n_{O2} + (O_2)_c n_{er} \} \qquad \cdots\cdots\cdots (11)$$

The quantity of the gas states in the combustion process can be calculated in six stages within a micro-time using Patterson's step method [10] with necessary modifications.

CONCENTRATION OF EXHAUST EMISSION GASES IN THE SEPARATED EXHAUST PIPE —— If what corresponds to η_{tr} at each levels of the separated exhaust pipe is given as η_{tri} (i = U, M or L, i.e. the upper, middle or lower level), the concentration of such exhaust gases, for instance, O_2 in an i-level portion of exhaust pipe can be determined by the following equation:

$$(O_2)_{ei} = [\eta_{tri} \{ n_{O2} + (O_2)_c n_{er} \}$$
$$+ (1 - \eta_{tri}) \frac{\gamma}{\gamma_0} (m + \frac{n}{4})]$$
$$/ [n_{fa} (\eta_{tri} \phi + 1 - \eta_{tri})]$$
$$\cdots\cdots\cdots\cdots (12)$$

COMBUSTION IN THE BURNER AND REACTION IN THE CATALYZER —— Let us assume that $\{(\gamma_{()} - \gamma_2)/\gamma_0\} (m + n/4)$ moles of secondary air are added to the exhaust gas from the two-stroke cycle engine so the air-fuel ratios in the burner and the catalyzer will be γ_B and $\gamma_C (= \gamma_{()})$. Where, $\gamma_{()}$ is shown as a representative of the air-fuel mixture ratio γ_B in the combustion of the after-burner and γ_C in the reaction of the catalyzer. If the fraction of the short-circuited unburned

mixture from the two-stroke cycle engine is given as $\delta = (1 - \eta_{tri})$ and exhaust gas from the engine is in a frozen state, the reaction equation under these condition can be formulated as follows using the unburned gas reduction rate η_{HC}:

$$\delta \{ C_m H_n + \frac{\gamma_2}{\gamma_0} (m + \frac{n}{4}) O_2$$
$$+ \frac{79.11}{20.99} \frac{\gamma_2}{\gamma_0} (m + \frac{n}{4}) N_2 \} + (1 - \delta) \{ n_{CO2} CO_2$$
$$+ n_{H2} H_2 + (n_{CO} CO + (CO)_c n_{er} CO) + n_{H2O} H_2O$$
$$+ (n_{O2} O_2 + (O_2)_c n_{er} O_2) + n_{N2} N_2 + n_{NO} NO \}$$
$$+ \frac{\gamma_0 - \gamma_2}{\gamma_0} (m + \frac{n}{4}) \{ O_2 + \frac{79.11}{20.99} N_2 \}$$
$$= (1 - \eta_{HC}) \delta C_m H_n + n'_{CO} CO + n'_{CO2} CO_2$$
$$+ n'_{H2} H_2 + n'_{O2} O_2 + n'_{H2O} H_2O + n'_{N2} N_2$$
$$+ n'_{NO} NO + n'_{OH} OH + n'_O O + n'_H H$$
$$+ n'_N N + (CO)_c n'_{er} CO + (O_2)_c n'_{er} O_2$$
$$\cdots\cdots\cdots\cdots (13)$$

The term $n'_{()}$ of the right hand side in the equation (13) represents the sum of the burned gas in the cylinder and the combustion gas of the short-circuited mixture. Furthermore, in the comparison analysis for NO formation by supposing two different chemical reaction mechanisms and evaluating the reaction kinetics rates, either of the reaction mechanisms expressed by equations of (5) or (7) was used.

The mole number of the compositions of gas in equation (13) along with the temperature, can be calculated as follows using the value of δ determined by the measurement of η_{tri}, the measured result of η_{HC} and based on changes in enthalpy Q which is supposed to be in the atmospheric state (P_0, T_0). Therefore, the emission reduction rate η_{CO}, η_{NO} for CO and NO emission can be calculated by the following equations:

$$\eta_{CO} = 1 - \frac{n_{CO}}{(1 - \delta) \{ n_{CO} + (CO)_c n_{er} \}}$$
$$\eta_{NO} = 1 - \frac{n_{NO}}{(1 - \delta) \, n_{NO}}$$
$$\cdots\cdots\cdots (14)$$

ENTHALPY AND EXERGY FLOW —— Figure 1 exhibits the result estimated from simulating the enthalpies $Q_{()}$ [J/s] and the exergy flows $E_{()}$ [J/s] in the two-stroke cycle engine, after-burner, catalyzer, Stirling engine and hot-water supply system.

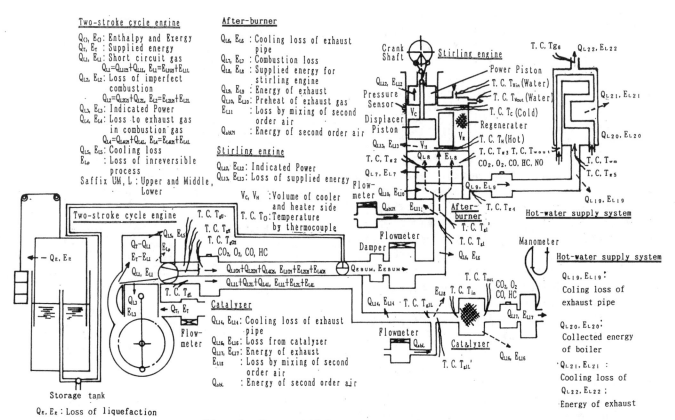

Two-stroke cycle engine

Q_O, E_O: Enthalpy and Exergy
Q_T, E_T: Supplied energy
Q_{LI}, E_{LI}: Short circuit gas
$\qquad Q_{LI}=Q_{LIUM}+Q_{LIL}, E_{LI}=E_{LIUM}+E_{LIL}$
Q_{L2}, E_{L2}: Loss of imperfect
$\qquad$ combustion
$\qquad Q_{L2}=Q_{L2UM}+Q_{L2L}, E_{L2}=E_{L2UM}+E_{L2L}$
Q_{L3}, E_{L3}: Indicated Power
Q_{L4}, E_{L4}: Loss to exhaust gas
$\qquad$ in combustion gas
$\qquad Q_{L4}=Q_{L4UM}+Q_{L4L}, E_{L4}=E_{L4UM}+E_{L4L}$
Q_{L5}, E_{L5}: Cooling loss
$E_{L\rho}$: Loss of irreversible
$\qquad$ process
Saffix UM, L: Upper and Middle,
$\qquad\qquad$ Lower

After-burner

Q_{L6}, E_{L6}: Cooling loss of exhaust
$\qquad$ pipe
Q_{L7}, E_{L7}: Combustion loss
Q_{L8}, E_{L8}: Supplied energy for
$\qquad$ stirling engine
Q_{L9}, E_{L9}: Energy of exhaust
Q_{L10}, E_{L10}: Preheat of exhaust gas
E_{L11}: Loss by mixing of second
$\qquad$ order air
Q_{aUM}: Energy of second order air

Stirling engine

Q_{L12}, E_{L12}: Indicated Power
Q_{L13}, E_{L13}: Loss of supplied energy
V_C, V_H: Volume of cooler
$\qquad$ and heater side
$T.C. T_0$: Temperature
$\qquad$ by thermocouple

Catalyzer

Q_{L14}, E_{L14}: Cooling loss of exhaust
$\qquad$ pipe
Q_{L16}, E_{L16}: Loss from catalyzer
Q_{L17}, E_{L17}: Energy of exhaust
E_{L18}: Loss by mixing of second
$\qquad$ order air
Q_{aL}: Energy of second order air

Hot-water supply system

Q_{L19}, E_{L19}:
Coling loss of
exhaust pipe

Q_{L20}, E_{L20}:
Collected energy
of boiler

Q_{L21}, E_{L21}:
Cooling loss of

Q_{L22}, E_{L22}:
Energy of exhaust

Q_E, E_E: Loss of liquefaction

Fig. 1 Energy flow and measurement
(gas temperature, composition and amount)

More specifically, the diagram indicates that most of the energy supplied to the two-stroke cycle engine is converted into engine output and a cooling loss, while the remaining heat energy as exhaust gas heat is discharged from the exhaust port. The exhaust gas contained the heat energy to be discharged is divided at the exhaust port into three parts: the upper, middle, and lower portions, of which the first two portions are combined and burned in a separately installed after-burner. The portion of the exhaust gas discharged from the lower section of the exhaust port is treated with a catalyzer and released into the environment. The diagram exhibits in addition to the energy flow the temperatures $T_{()}$ and the measuring points for the gas compositions (i.e. CO_2, O_2, CO, HC, and NO).

TEST EQUIPMENT AND METHOD

As in the previously reported experiment, a two-stroke cycle engine with the rotary exhaust valve and the exhaust port divided into three sections (i.e. upper, middle, and lower outlets) was used for the

test discussed in this paper. The method of exhaust gas separation and the separated gas treatment are briefly described in the following. The portions of exhaust gas discharged from the top and middle sections of the exhaust port are temporarily stored in a tank. And then, these portions of gas are burned a second time in an after-burner unit to recover the heat energy thus generated with the operation of a Stirling engine and a hot-water supply system. Since the portion of exhaust gas from the bottom section of the port contains burned gas in high concentration and HC in low concentration, it is conducted directly into a catalyzer to control it of the pollutants. The catalyzer is a three-way catalyst unit.

The two-stroke cycle and Stirling engines used for the test are the same equipment that were used for the previous experiment. The two-stroke cycle engine has a cylinder with a bore of 62 mm X 58 mm in stroke. The Stirling engine has a bore of 45 mm X 34 mm in stroke. The gas tank for temporal storage is a floating-bell type with a volume of 1,000 [ℓ]. The hot-water supply system is an ordinary market commercial product. The catalyzer, also a commercially available product, is a three-way catalyst for controlling such gas pollutants as CO, HC and NO.

The test was conducted with the carburetor at full throttle. The concentrations of O_2, CO, HC and NO were measured by a continuous gas analyzer at several points where there was gas flow, i.e. at the outlets of the two-stroke cycle engine, the after-burner, and the catalyzer. The measured consentrations of O_2 from the upper, middle, and bottom sections of the exhaust port were used for the calculation of the concentration of unburned short-circuited charge mixture $\delta = 1 - \eta_{tri}$ in the exhaust gas. As in the previous paper, the formulas developed by Miyabe [12] and Tomizuka [13] were employed to calculate η_{tri}. The temperatures and secondary air were also measured in the same way as in the previous paper.

As in the previous paper, this paper conducts the similar formula-handling of combustion in the cylinders for the generating mechanism of the emission gases CO, HC and NO giving adequate attention to the removal of these pollutants as well. This has of course a close bearing on the compression, combustion, and expansion strokes of the engine. An in-cylinder pressure indicator diagram was measured with an engine analyzer.

TEST RESULTS AND DISCUSSION

Figure 2 exhibits the experimentally measured and theoretical in-cylinder pressure/crank angle curves at an engine speed $n_2 = 2,500$ rpm and an air-fuel ratio $r_2 = 13$. Further conditions of measurement and calculation are as follows. At the beginning of the compression stroke $\alpha_{EC} = 110°$ BTDC (Exhaust Open $\alpha_{EO} = 110°$ ATDC), the in-cylinder gas temperature T_{ZA} was 500 K, the gas pressure P_{ZA} was 0.117 MPa, the cylinder wall temperature T_w was 490 K, piston temperature T_P was 550 K. The spark ignition timing of crank angle α_{IG} was 15 degrees BTDC, the crank angle at the maximum pressure α_{Pmax} was 15 degrees ATDC, and scavenging efficiency η_s was 0.64. In the combustion calculation, the crank angle at the maximum in-cylinder pressure was made to coincide with the end of the burning time, and the mass burning rate was corrected so

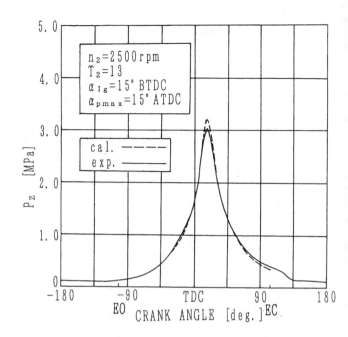

Fig. 2 Pressure diagram

that, at that time, its fraction should be 100%. The fuel used for the calculation was C_8H_{18}. As is apparent from Figure 2, the measured maximum pressure was slightly lower than the theoretical value because the end-point of the combustion extended well into the expansion stroke. It is safe to say, however, that the measured and calculated results for a full cycle coincide closely. The results of calculations based on the Zeldovich and bimolecular reaction mechanisms are indicated by one line as they do not differ widely enough to justify being presented separately in the figure scale.

Figure 3 summarizes the results of focusing on the formation of NO under the same conditions as given in Figure 2, except that the air-fuel ratio was varied in several ways. The figure exhibits, from top to bottom, the temperature of the burned gas T_B, the temperature of the unburned gas T_U, the oxygen concentration O_2, the in-cylinder pressure P_Z, and the concentration of NO. These data indicate that the formation of NO depends on the synergestic effect of T_B and O_2 concentration. In other words, NO formation depends largely on the temperature when it is in a fuel rich mixture range, whereas NO formation is significantly affected both by temperature and the O_2 content of the gas when it is in a fuel lean mixture range. The highest concentration of

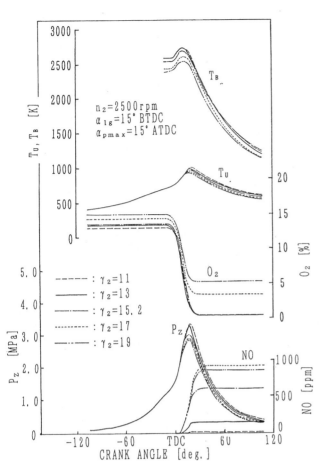

Fig. 3 Temperatures and gas concentrations

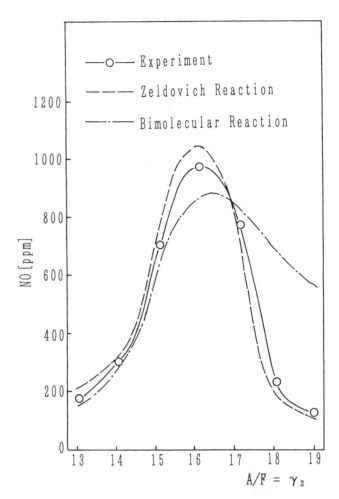

Fig. 4 Comparison with calculations and experiments for NO concentration

NO occurs around an air-fuel ratio of 16 or 17, because that ratio achieves the maximum synergestic effect of the two factors mentioned above. This figure is based on the bimolecular reaction mechanism, but corresponding data based on the Zeldovich reaction mechanism show the same tendency inrespect of NO formation according to time (crank angle).

Figure 4 compares the experimentally measured and theoretical (both Zeldovich and bimolecular reaction mechanism based) NO formation levels in exhaust pipe as recorded when the air-fuel mixture ratio γ ranged between 13 and 19. Both the measured and the calculated values reveal that the NO concentration climaxes at an air-fuel mixture ratio of 16, while it tends downward both in the lower and higher ranges of air-fuel mixture ratio. The NO concentration as measured on the basis of the Zeldovich reaction mechanism closely coincides with the measured and calculated values over a wide range of air-fuel mixture ratios.

By comparison with the bimolecular

reaction mechanism based calculation, the calculation based on the Zeldovich reaction mechanism indicates a decrease in NO concentration in the fuel-lean mixture range of air-fuel ratio. This implies that the concentration of this pollutant gas depends more on temperature than on oxygen level. It may therefore be inferred that the experimentally measured NO concentration will show a similar tendency in this respect.

Figure 5 exibits the temperatures T_{g1} and T_{g2} in the after-burner and the concentrations of O_2, CO_2, HC, and NO according to two operating conditions. Under one test condition, the two-stroke cycle engine was operated at an engine speed of $n_2 = 2,500$ rpm and an air-fuel charge mixture ratio of $\gamma_2 = 13$. Under this test condition the secondary air was added to the exhaust gas parts directly from the upper and middle sections of the exhaust port in order to vary the air-fuel ratio γ_B of mixing gas

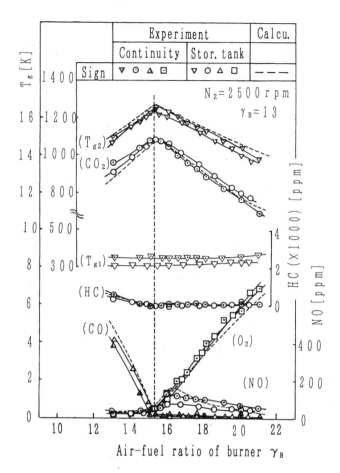

Fig. 5 Concentration and temperature of
exhaust gas in continuous operation
and storage tank use
(upper and middle portions)

the gas concentrations at that γ_B can be correspondingly ascertained. In this case it is possible to find out the temperature T_{g2} at the after-burner outlet by determining the changes in enthalpy, as expounded in the previous paper [14]. Calculation is made from the temperature T_{g1} at the after-burner outlet by holding the pressure in this installation equal to the atmospheric pressure.

Figure 5 shows that the experimentally measured and calculated results coincide well. The experimental measurement and the calculation of both the temperatures and concentrations of gas coincide closely both during continuous operation and storage in the charging tank. They also agree closely with the relevant theoretical values. The diagram indicates that the NO concentration in the charging tank is somewhat lower than in operation because some of the NO content of the exhaust gas is absorbed by the water in the tank.

Figure 6, a counterpart of figure 5, shows how the catalyst conversion efficiency rates η_{CO}, η_{NO}, and η_{HC} vary with γ_B. The conversion efficiency η_{CO} and η_{HC}, for example, are almost perfect above the stoichiometric air-fuel mixture ratio of the

into the after-burner. Under the second test condition, the engine was operated at the same engine operating condition, but the exhaust gas from the same outlets was first stored in a storage tank before the secondary air was added to it. The figure also exhibits the calculated results as based on the Zeldovich reaction mechanism.

The following parameter values are substituted into equation (13). $\gamma_{()} = \gamma_2 = 13$, that is, no secondary air is added to the exhaust gas. Furthermore, m = 8, n = 18, and $\delta = 1 - \eta_{trum} = 0.52$ in measured value. Finally, the gas compositions in frozen equilibrium n'_{CO}, n'_{H2}, ... and $\eta_{HC} = 0.97$ in measured value. Under these premises the values of n'_{O2}, n'_{H2O}, ... can be determined and the concentration of each gas composition can be accordingly calculated. If there is a change in $\gamma_{()} = \gamma_B$, and only η_{HC} is varied in measured value, the values of n'_{O2}, n'_{H2O}, ... and

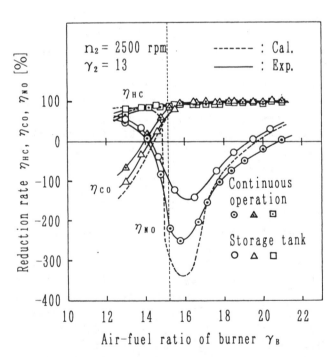

Fig. 6 Reduction rate of emission gas in
continuous operation and storage
tank use
(upper and middle portions)

after-burner, whereas η_{NO} falls to its lowest level around an air-fuel mixture ratio of 16 or 17. The calculated and measured results nearly coincide both in continuous operation and storage tank and they approximate the theoretical by calculated values.

In the window of catalyst operation in which NO, CO, and HC are reduced simultaneously [15], a slight variation in the air-fuel mixture ratio produces a remarkable change in the conversion efficiency of CO and NO.

Figure 7 exhibits both experimentally measured and calculated concentrations and temperatures of emission gas components when exhaust gas from the lower section of the two-stroke cycle engine's exhaust port was treated by a three-way catalyzer. Substituting $\delta = 1 - \eta_{trL}$ and $\eta_{trL} = 0.86$, the results were calculated by the same method as in the case of exhaust gas supplied to the after-burner. As also is apparent from the figure, the measured and calculated results coincide well. A Zeldovich reaction was used to ascertain the NO formation in

this calculation as well.

Figure 8 shows the conversion efficiency η_{CO}, η_{NO}, and η_{HC} that correspond to Figure 7. In this case also, the similar trend as shown in Figure 6 is indicated. For instance, above the stoichiometric air-fuel ratio of mixture into the catalyzer, η_{CO} and η_{HC} are approximately 100 % corresponding almost complete conversion efficiency, while maximum reducing efficiency η_{NO} are achieved around the air-fuel ratio of 16 to 17.

Figure 9 shows the enthalpy flow when the two-stroke cycle engine was operated at a engine speed of $n_2 = 2,500$ rpm and at an air-fuel ratio of charge mixture $\gamma_2 = 13$. $\gamma_2 = 13$ is almost the optimum value of average effective pressure. Secondary air was added to the exhaust gas from the engine to ensure that the air-fuel ratios during the reactions in the after-burner and catalyzer would be γ_B, $\gamma_C = 20$. The upper left corner of figure shows the measurement of the enthalpy flow when a storage tank for a part of exhaust gas is used.

Figure 10 shows the exergy flow as determined under the same conditions as given for Figure 9.

The transition of flow given in these two figures can be corresponded to Figure 1. Nearly the same method for calculating the energy was used as in the previous paper.

Of the total energy supply to the two-stroke cycle engine, $(29.55 + 5.52)$ % $(Q_{LIUM} + Q_{LIL})$ in enthalpy, and $(32.15 + 6.55)$ %

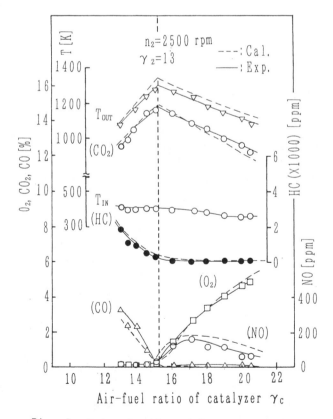

Fig. 7 Concentration and temperature of exhaust gas in a catalyzer use (lower portion)

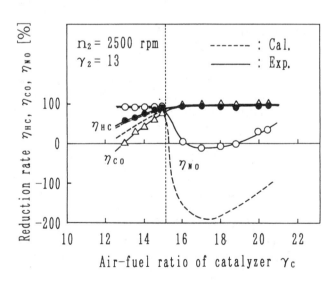

Fig. 8 Reduction rate of emission gas in a catalyzer use (lower portion)

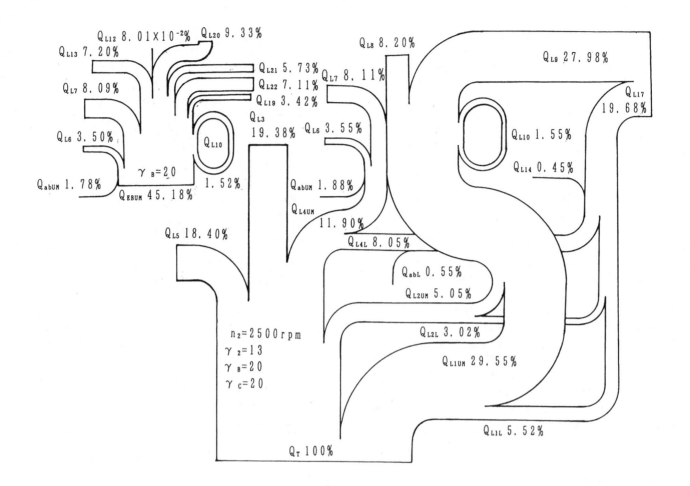

Fig. 9 Enthalpy flow

$(E_{L1UM} + E_{L1L})$ in exergy are lost as an unburned short-circuited charge mixture through the exhaust port. Of this energy loss, the portion short-circuited through the upper and middle sections of the exhaust port amounts to 29.55 % (Q_{L1UM}) in enthalpy and 32.15 % (E_{LUM}) in exergy. Considering that the work performed by the two-stroke cycle engine is 19.38 % (Q_{LS}) in enthalpy and 20.63 % (E_{LS}) in exergy, the recovery of energy loss indicated above suggests that the system proposed by this paper would be very useful.

This is strongly urges by the fact that an enthalpy of (7.20 + 0.08) % ($Q_{L13} + Q_{L12}$ = Q_{L8}) + 9.33 % (Q_{L20}) and an exergy of (6.57 + 0.079) % ($E_{L13} + E_{L12} = E_{L8}$) + 3.10 % ($E_{L20}$) can be extracted from exhaust gas by burning it again in a after-burner unit. Incidentally, available energy declines somewhat more in using a storage tank than in continuous operation. The energy at the after-burner outlet Q_{L9}, for instance, is

27.98% in the former case, compared with 25.59% in the latter. From these results, one can verify obviously the possibility that the most heat energy to be lost can be stored temporarily as a form of exhaust gas at a storage tank and then can be recovered.

CONCLUSION

The findings of the study discussed in the preceding sections may be summarized as follows:

(1) The study demonstrates that exhaust gas containing an unburned mixture in high concentration from a exhaust gas-separate type two-stroke cycle engine can be temporarily stored in a storage tank. A comparative analysis of measured exhaust gas concentrations and energy-flow in the storage process and in continuous operation supports the effectiveness of the storage process.

(2) A Stirling engine was able to recover

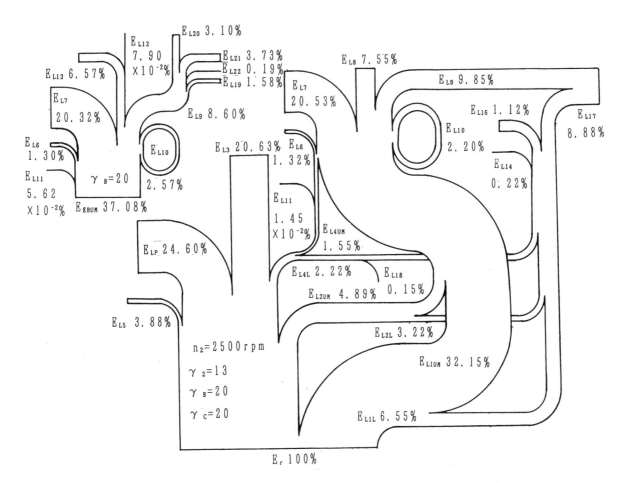

Fig. 10 Exergy flow

exhaust heat energy by after-burning the exhaust gas after it was once stored in a storage tank and furthermore, a hot-water supply unit added newly could be operated by remaining heat energy after operating the Stirling engine. It is thus verified more cogently than in the previous paper that the proposed system effectively utilizes exhaust gas.

(3) The exhaust gas-separate type two-stroke cycle engine, can divide exhaust gas into a portion with high concentration of unburned charge mixture and another that contains burned gas in high concentration. By the catalyzer treatment of the former portion after its effective heat utilization and of the latter immediately after its discharge from the two-stroke cycle engine, nearly 100% of the CO and HC was removed from the emission gas. Even in a lean air-fuel mixture ratio range at present report, the NO content was also reduced

remarkably, as expounded in the previous paper.

(4) Dissociation reactions were used to simulate the formation of such emission gas components as CO, HC, and NO through fuel combustion in the two-stroke cycle engine. The bimolecular and Zeldovich reaction mechanisms in particular were used to investigate NO formation. The findings from the simulation suggest that the separation process depicted above purifies exhaust gas of these components to a satisfactory degree. The application of Zeldovich reaction mechanism is a more accurate means of verifying the NO content of exhaust gas.

NOMENCLATURE

Cp: Specific heat in constant pressure

E: Exergy
   The maximum external work that the gas at a pressure P and temperature T can

perform for atmospheric condition (Po, To), i.e. the exergy, may be expressed as follows:

$$E = G \{ C_P (T - T_0) - C_P \ell n \frac{T}{T_0} + RT_0 \ell n \frac{P}{P_0}$$

- G: Amount of Gas
- G_{ab}: Amount of secondary air
- Po: Atmospheric pressure
- Q: Enthalpy
- R: Gas constant
- r_r: Participation rate in the combustion of residual gas
- To: Atmospheric temperature
- γ: Actual air-fuel mixture ratio
- γ_0: Theoretical air-fuel mixture ratio
- δ: Concentration of short-circuited gas
- η_{CO}: Conversion efficiency of CO
- η_{HC}: Conversion efficiency of HC
- η_{NO}: Conversion efficiency of NO
- η_s: Scavenging efficiency
- η_{tr}: Trapping efficiency
- ϕ: Dry exhaust gas volume fraction to charge mixture

ACKNOWLEDGMENT

We wish to thank Sasamura Engineering Promotion Association and postgraduates and undergraduates of Shibaura Institute of Technology for their valuable assistance and cooperation in carrying out this study.

REFERENCES

[1] Sato, K., Ukawa, H., and Nakano, M., "Effective Energy Utilization and Emission Reduction of the Exhaust Gas in a Two-Stroke Cycle Engine," SAE Paper, 911848 (1991).

[2] Sato, k., and Nakano, M., "Separation of Exhaust Gas in Two-Stroke Cycle Gasoline Engines," Transaction of Japan Soc. of Mech. Engrs., 52-482, B(1986), PP. 3616.

[3] Sato, K., and Nakano, M., "A Method of Separating Short-Circuit Gas from Exhaust Gas in Two-Stroke Cycle Gasoline Engines," SAE Paper, 871653 (1987).

[4] Sato, K., and Nakano, M., "Separation of Exhaust Gas in Two-Stroke Cycle Gasoline Engines (a follow-up)," Transaction of Japan Soc. of Mech. Engrs., 54-497, B(1988), PP. 197.

[5] Sato, K., Nakano, M., and Ito, Y., "Separation and Effective Utilization of Exhaust Gas in Two-Stroke Cycle Engines," Transaction of Japan Soc. of Mech. Engrs., 56-526, B(1990), PP. 1843.

[6] Sato, K., Ito, Y., and Nakano, M., "Effective Energy Utilization and Emission Reduction of Exhaust Gas in Two-Stroke Cycle Engines," Transaction of Japan Soc. of Mech. Engrs., 57-534, B(1991), PP. 756.

[7] Sato, K., Ukawa, H., and Nakano, M., "Effective Energy Utilization and Emission Reduction of the Exhaust Gas in a Two-Stroke Cycle Engine," SAE Paper, 911848 (1991).

[8] Kaufmann, F., and Kelso, J. K., "Thermal Decomposition of Nitric Oxide," J. J. Chem. Physics, 23-9 (1955), PP. 1702.

[9] Newhall, H. K., "International Comb. Conf. Twelfth Symp. (International), on Combustion, (1969), PP. 603.

[10] Sato, K., Kido, K., and Nakano, M., "A Method on Improvement of a Two-Stroke Cycle Gasoline Engine for General Purpose and Agricultural Use. (2nd Report)," Journal of Agricultural and Mechanical Engineering Society, 41-2 (1977), PP. 193.

[11] Patterson, D. T., "A Digital Computer Simulation for Spark-Ignited Engine Cycles," SAE Trans., 14-18 (1963), PP. 633.

[12] Miyabe, H., and Shimomura, R., "A Study on Trapping Efficiency of Two-Stroke Cycle Gasoline Engines," Transaction of Japan Soc. of Mech. Engrs., 28-192 (1962), PP. 1006.

[13] Tomizuka, K., and Shibata, H., "On Method of Measuring Charging Efficiency of Two-Stroke Cycle Engine," Aeronoutical Research Institute report, 61 (1929), PP. 455.

[14] Sato, K., Ukawa, H., and Nakano, M., "Effective Energy Utilization and Emission Reduction of the Exhaust Gas in a Two-Stroke Cycle Engine," SAE Paper, 911848 (1991).

[15] Tada, A., "Combustion and Exhaust Emission in Automotive Engines," Journal of Internal Combustion Engine (Japan), 29-366 (1990), PP. 65.

Preignition with High Olefin Fuels in a 2-Stroke SI Engine

R. L. Mendiratta and B. P. Pundir
Indian Institute of Petroleum

ABSTRACT

Hydrocarbon composition of fuel affects the deposit composition, its capacity to heat up the hot spots, and propensity of the fuel to preignition. Presently, fluidized catalytically cracked streams forms a large fraction of total gasoline pool in India and gasolines contain up to 50% olefins. About 60% of total gasoline in the country is consumed by the two wheeled vehicles powered mostly by 2-stroke engines. Preignition tendency of fuels with varying content of olefinic hydrocarbons was studied on a 2-stroke engine, during a 50 hour test. Preignition was indicated by sudden increase in combustion chamber surface temperature. Results showed a marked increase in preignition as the olefin content of gasoline increased above 20% by volume.

INTRODUCTION

Preignition is an uncontrolled surface ignition of the carburetted fuel-air mixture by hot spots or by glowing combustion chamber deposits prior to the normal spark ignition. It is generally observed in high compression gasoline engines during high load operation after a certain critical thickness of deposits has been formed in the combustion chamber. The consequences of the preignition include, loss of engine power, increase in engine roughness and noise, overheating of engine parts and failure of engine structure such as blowing hole in the piston and breaking piston rings and lands, etc.

Air-cooled, high speed and high output two-stroke gasoline engines may be more prone to preignition than 4-stroke gasoline engines mainly because of the following factors,

-Poor scavenging and absence of internal cooling during the induction stroke,

-Air cooling leading to more thermal loading, and

-Total oil loss system resulting in more combustion chamber deposits.

Among other factors, the preignition depends upon the deposit formation tendency of fuels and lubricants. Various studies (1-4)*on 4-stroke gasoline engines have shown that the hydrocarbon composition of fuel not only affects deposit formation tendency but also capacity of fuel to heat up the

----

( )* Number in parentheses designate to references at the end of the paper.

hot spots in the combustion chamber as well as the propensity of fuel to preignite. Very little work on the effect of fuel composition on preignition in 2-stroke engines is reported, although 2-stroke engine lubricants and the nature of their additives are well known to have a significant influence on preignition.

In India, the two wheelers viz., mopeds, scooters and motorcycles are the principal means of individual transport instead of cars and their population is comparatively much larger. The trends in the growth of population of these vehicles are given in Table - 1 (5). This pattern of vehicle population has resulted in about 60 percent of total gasoline being consumed by the two and three wheelers annually in the country. The passenger cars are invariably powered by 4-stroke gasoline engines while the two and three wheelers are mostly powered by air-cooled, two-stroke gasoline engines of 50 to 175 cm3 displacement.

As regards the gasoline refining, out of a total of 12 refineries, 7 refineries have fluidized catalytic cracker (FCC) units. During 1991-92, total gasoline production was 3.8 million tonnes, out of which 3.2 million tonnes was produced by the refineries equipped with FCC units. In these refineries, FCC gasoline constituted 45 to 68% of the gasoline pool and the gasoline contains upto 50% olefins. Earlier studies (1-4) on 4-stroke SI engines, have shown a high propensity of olefins to cause preignition. The air-cooled, two stroke engines being otherwise more prone to preignition, the problem may be further aggravated if highly olefinic gasolines are used.

In view of the above, in the present work preignition tendency of fuels with varying content of olefins was studied on an air-cooled, 2-stroke engine which

is used in large numbers to power two wheeled vehicles in India. In addition, the effect of a few different oil formulations and oil/fuel ratio were studied. In this paper findings of these studies are discussed.

Table -1
Population of Passenger Cars and
2-Wheeler Vehicles in India(5)
,000's

Year	Passenger Cars	2-Wheelers
1971	682	576
1975	766	946
1980	1055	2115
1985	1546	5121
1990	2317	13034
1992 (estimated)	2766	16936

LITERATURE REVIEW

The fuel, lubricant and air are the main source of engine deposits. Due to incomplete combustion of fuel and lubricant, carbonaceous deposits containing various metal oxides and salts originating from the fuel and lubricant additives are formed which may lead to preignition. The composition of combustion chamber deposits depends upon the physico-chemical nature of fuels, lubricants and their additives and engine operating conditions.

Felt, et al.(1), studied the effect of fuel hydrocarbon composition on surface ignition tendency in a multicylinder, 4-stroke gasoline engine having 10:1 compression ratio (CR) by recording number of times the test fuel was ignited by hot spots in each cylinder. When the olefin content of fuel increased from 0 to 30% , the surface ignition count increased from 53 to 76.Further increase in olefin content to 50% increased preignition count sharply ranging from 121 to 146. They established that there is a fair degree of correlation between the surface ignition tendency and olefin content of the fuel. Among aromatics, benzene was found to be highly prone to preignition.

Guibet, et al.(2), in their studies on a single cylinder, 380 cm3, 9.4 CR, 4-stroke engine demonstrated that the fuel effects on preignition depend upon its resistance to hot spot ignition and capacity to heat up the ignition source. The latter is linked to the flame speed and flame temperature characteristics of the fuel. Paraffinic hydro-carbons have relatively low carbon to hydrogen ratio and hence lower adiabatic flame temperatures than aromatic hydrocarbons and so cause less heating of combustion chamber deposits. For example xylene, although it has lower flame speed than iso-octane, still heats up the deposits and other surface ignition source to a greater degree mainly because of its higher flame temperature. They further, observed that toluene and xylene being highly resistant while cyclohexane and di-isobutene were very prone to preignition, thus underlining the complex behaviour of fuels towards surface ignition. Addition of tetra ethyl-lead (TEL) increased the temperature of deposits when used with less resistant hydrocarbons such as olefins. In general, these investigators also observed a satisfactory performance of paraffins and isoparaffins but low resistance of olefins They further, concluded that a mixture of two hydrocarbons offer less resistance than expected by simple linear interpolation between the properties of the constituents.

Holger Menrad, et al.(3), also observed paraffins to have high resistance to preignition unlike napthenes, especially cyclohexane. Aromatics showed no clear trend. These authors in their studies on a 4-stroke, multicylinder engine found that the addition of 50% isopentane to methanol (amongst lowest on the preignition scale) reduces the preignition susceptibility of methanol significantly. These authors concluded that no correlation appeared to exist between research or motor octane numbers and preignition susceptibility.

Lubricant additive composition has a profound effect on pre-ignition occurrence, particularly in two stroke engines. Gow, et al. (6) studied the effect of metal content of oil on preignition in outboard 2-stroke engine tests. Preignition increased as the metal content of the oil fuel blend increased. Ca and Ba had predominant effect on preignition while Mg additives showed no tendency to cause preignition (6,7). Pless (7) reported the relative preignition propensity of some metals used in Lubricants. With respect to Ba, which was assigned a rating equal to 1.0, Mg, K and Ca had a rating of 0.32, 1.42 and 2.6, respectively. Zinc had little effect while phosphorous and sulphur were observed to reduce preignition.

Guibet, et al.(2), observed that time required for preignition occurrence varied inversely with additive concentration in the lubricant. Ca and Ba again were predominant in causing preignition which occurred after almost constant consumption by mass of combined Ba and Ca additives. Preignition occurred just after 60 hours when oil consumption was 1% of fuel consumption compared to 200 hours with 0.25% oil consumption. It was observed that the operating time before the onset of preignition was lowered considerably when the lubricant consumption was increased. Gow, et al.(6), in their studies on an outboard 2-stroke engine found that while preignition was observed with 5% oil/fuel ratio, but with 1% oil/fuel ratio, no preignition was observed

It is also reported(9-12) that ashless additives in 2T oil reduce the tendency of deposit formation and are less likely to cause preignition, However, some of these additives are less effective in reducing piston deposits and bearing corrosion than the metal based additives.

Phosphorus additives e.g. tricresyl phosphate have been used earlier to reduce surface ignition tendency of leaded gasoline in high compression ratio engines. Recently, Bert, et al.(7),have reported reduction in run on tendency of car with an additive concentrate based on polyether amine chemistry of dispersant type. This additive reduced combustion chamber deposits, thus reducing run-on. In their tests with unleaded gasoline of a 5.0 L V-8 laboratory engine and a 30-car fleet, using a procedure adopted from the chassis dynamometer, run on tendency was reduced in 63% of the cars. This was attributed to efficacy of the additive in reducing combustion chamber deposits.

PREIGNITION MEASUREMENT TECHNIQUES

Different preignition measurement techniques are reported in the literature, viz.,

(i) Ionization detector method was used by Felt, et al.(1), and Guibet, et al.(2). This method consists of interrupting spark ignition for a very brief moment and detection of eventual surface ignition by means of an ionization detector located in the combustion chamber on an oscillograph (1,2). At the occurrence of preignition the appearance of ionization signals are accompanied by a sharp decrease in engine power. The degree of preignition is estimated by crank angle at which ionization is detected.

(ii) Tomsic (4) used combustion pressure time trace in their study of surface ignition behavior of fuels. Surface ignition causes peak combustion pressure to occur earlier than normal in the combustion cycle. The magnitude of the advance in time of peak pressure in combustion cycle was taken as a measure of severity of surface ignition.

(iii) The Coordinating Research Council (CRC) originally proposed a leaded iso-octane and benzene (LIB) reference fuel system for measurement of surface ignition resistance of fuels similar to that used to measure the octane number of fuels with the iso-octane and n-haptane system (9). Isooctane has high surface ignition resistance relative to benzene and the blends of the two are taken to have a varying intrinsic resistance to deposit ignition for comparison with the surface ignition characteristics of the sample fuel.The LIB number denotes the percentage of iso-octane in the blend. Higher the LIB number, higher is the resistance to surface ignition of a fuel.

(iv) Several test methods have been developed to study the preignition tendency of engine oil formulations. These are generally long duration tests comprising a deposit build-up phase followed by high load operation. In the latter phase preignition occurs if lubricant additives are not suitable. The Fiat 132C preignition test(14) uses a four cylinder, 4-stroke SI engine and cyclic operating conditions during the deposit build-up phase The fuel is mixed with 1% by vol. of test oil. Preignition is detected by sudden increase of combustion chamber wall temperature above the normal operating value measured by a thermocouple attached to the spark plug seat. This is accompanied by loss of engine power of about 30%. The performance of the test oil is evaluated by the engine running hours before preignition is observed.

(v) The Vespa 2-stroke engine preignition test (15) is intended for the evaluation of 2-stroke engine lubricants for deposit induced preignition. The test is run on a Vespa 180 SS engine at 4000 rpm under wide open throttle (WOT) conditions. The onset of preignition is detected by sudden increase in combustion chamber surface temperature measured by a shielded thermocouple mounted in the cylinder head accompanied by sudden engine power loss. After each incidence of major preignition, the engine combustion chamber and exhaust port deposits are cleaned and the test is continued. The number of preignition observed during the 50-h test is the measure of severity of preignition with the test oil.

EXPERIMENTAL STUDIES

TEST METHOD-In the present study, a test method similar to Vespa 180 SS test procedure was developed on Bajaj Super 150 cm^3 2-stroke engine (16). The engine specifications are given in Table-2 and the test operating conditions are given in Appendix-I. The combustion chamber surface temperature is measured by a shielded iron-constanton thermocouple mounted in the cylinder head flush with the combustion chamber surface. The installation of the thermocouple is shown schematically in Fig.1. The spark plug seat temperature was mainttained at 190+5^{0}C by regulating cooling air to the cylinder block. A multi-channel recorder is used for continuous recording of combustion chamber surface and spark plug seat temperature and engine torque. To increase the test severity, the engine spark timing is advanced to 25^{0}btdc from the standard timing of 22^{0}btdc.

When the cylinder head temperature increases rapidly by 30^{0}C or more above its stable tempera-

ture recorded during 90 minutes of stabilization period it is taken as an indication of the major preignition. The incidence of preignition,is further evidenced by sudden increase in spark plug seat temperature and drop in the engine torque. A typical record of the spark plug seat and combustion chamber surface tempeeratures, along with the engine torque at the time of preignition are shown in Fig-2.

At each incidence of major preignition the engine is stopped promptly by a relay control circuit to prevent structural failure of the engine. Engine combustion chamber and exhaust port deposits are then cleaned and the test is continued with a new spark plug. The number of preignitions observed during 50 hour

Table - 2
Engine Specifications
Single cylinder 2-stroke forced air cooled, spark ignition gasoline engine

Bore, mm	57
Stroke, mm	57
Capacity, cm^3	145.45
Compression ratio	7.4:1
Max. power at 5200 rpm, kW	4.5
Ignition timing, ^{0}CA btdc	25$\pm$1

Thermocouple Hole

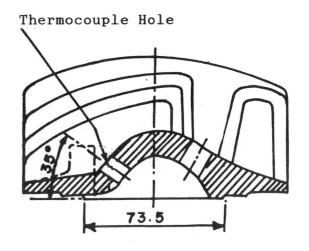

Fig.1: Combustion Chamber Thermocouple Location In Cylinder Head

test is the measure of pre-ignition tendency of the test fuel or oil. To study the effect of fuel quality, an oil which repeatedly gave no preignition with the reference fuel was used. TEST FUELS - The following test fuels were used to study the effect of olefin content of gasoline on preignition,

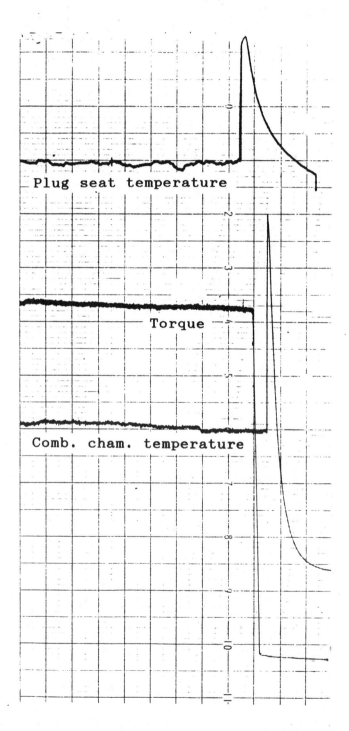

Fig.2: Typical record of combustion chamber and spark plug seat temperatures and engine torque at the onset of preignition.

Fuel	Description
A	A reference fuel
B	Reformate
C	FCC gasoline
D	50% FCC+50% Reformate blend
E	30% FCC+50% Reformate blend
F	FCC gasoline+1500 ppm of a multifunctional additive (MFA).

Physico-chemical characteristics of the test fuels are given in Table-3. These fuels were mixed with 5% by volume of a 2-stroke engine oil 'A' which had shown 'No Preignition' tendency with the reference fuel having 18.3% olefins. To increase the test severity lead content of fuel was maintained at 0.5 g Pb/litre.

RESULTS AND DISCUSSION

The discriminating ability of this test procedure was first established with different oil formulations. The oils used were CEC low reference oil RL81, high reference oil RL05, a good performance marketed oil formulation A, and a crankcase oil formulation B. The physico-chemical characteristics of these lubricants are given in Table-4. Preignition results obtained with the above oils using reference gasoline are shown in Table-5. With RL-81 five number of preignitions were observed. Oil B which had 0.44% sulphated ash gave 2 preignitions while low ash lubricants RL05 and Oil A gave no preignition. To evaluate the effect of fuel olefin content subsequent tests were done using oil A alone.
EFFECT OF GASOLINE OLEFIN CONTENT -Results on preignition obtained with the different test fuels during 50 h test duration are given in Table-6. With 100% FCC gasoline (olefin content 44.8%) 2 to 3 major preignitions were observed. When the oil content in fuel was reduced from 5% to 2%, no change in pre-ignition occurrence was observed as still 3 numbers of preignition were observed. With 50:50 blend of FCC gasoline and reformate,

Table-3
Physico-chemical Characteristics of Test Fuels

Characteristics	Fuel					
	A	B	C	D	E	F
Relative density at 15.6⁰C	0.739	0.743	0.715	0.732	-	-
Reid vapour pressure(RVP),kg/cm²	0.56	0.51	0.47	0.498	-	-
Existent gum , mg/100 ml	-	2.2	7.0	2.3	2.7	-
Potential gum, mg/100 ml	-	28.9	82.2	55.5	34.9	73.4
Copper corrosion	<1a	-	-	<1a	-	-
Sulphur, % mass	0.12	-	0.2	-	-	-
Lead, g/l	0.48	Nil	0.5	0.5	0.5	-
Distillation, ⁰C,						
IBP	38	32	40	31	41	-
10%	55	62	55	55	63	-
50%	87	98	88	90	94	-
90%	112	135	226	135	130	-
FBP	141	169	160	163	171	-
HC composition, %						
Saturates	33.8	57	44.3	50.6	53.2	-
Olefins	18.3	1.0	44.8	22.9	14.1	-
Aromatics	47.9	42	10.9	26.5	32.7	-

Table-4
Physico-chemical Characteristics of Test Oils

Characteristics	Test Oils			
	RL 81	RL 05	A	B
Sulphated Ash, % mass	1.29	0.25	0.13	0.44
Elemental Analysis, % mass				
(a) Barium	0.06	Nil	Nil	-
(b) Calcium	0.26	Nil	0.05	0.078
(c) Zinc	0.08	0.09	Nil	0.086
(d) Sulphur	0.68	-	0.86	0.85
(e) Phosphorous	0.08	0.09	Nil	0.08
(f) Nitrogen	0.02	0.05	0.018	-

preignition occurred once only. This blend had olefin content of 22.9%. However, with 30% FCC+70% reformate blend (olefin content 14.1%) and reformate alone no preignition was observed. Similarly with reference fuel (olefin content 18.3%), no preignition was observed. In the present study, frequency of unacceptable quantum of exhaust port deposit build up also increased with the gasoline having high olefin content. However, as the tests were not designed specifically for this purpose, no definite

Table - 5
Preignition Test Results with Different Oils

Oil	Oil/Fuel ratio, % vol.	Preign. at test hours	Numbers of Preign.
RL 81	5	8.5, 13, 21.5 28.5, 43	5
RL 05	5	-	Nil
A	5	-	Nil
B	5	7, 41.5	2

Table - 6
Results of Preignition Tests
(Test Duration, 50 h.)

Fuel	Oil/Fuel ratio, % vol	Exh. port cleaned* No. of times	Preign. at test hours	Numbers of Preign.
A	5	2	-	Nil
B	5	Nil	-	Nil
C	5	2	8,20.5,30	3
	5	2	20,47	2
	2	2	9.5,18.75,47	3
D	5	Nil	21	1
E	5	Nil	-	Nil

* Other than those at occurrence of preignition when the engine power decreased to 90% of its initial valve.

Table - 7
Preignition Susceptibility of Test Fuels

Test sequence	Fuel	Oil/Fuel ratio % vol	Test duration h	Preign. at test hours	Numbers of Preign.
I	C	5	50	20,47	2
II	E	5	20	-	Nil
III	C	5	20	13.5	1

Note: During this test the combustion chamber deposits were not cleaned when switched over to the next test sequence.

conclusions are being drawn. Confirmation of the preignition susceptibility of FCC gasoline was further attempted by an additional test conducted in the following sequence :

I. A 50-h test with 100% FCC fuel was first run.

II. With the deposits intact, this was followed by engine operation for 20 hours on 30% FCC+70% Reformate blend.

III. Then, test was continued again on 100% FCC fuel for another 20 h.

Results are given in Table-7. In sequence I, second preignition occurred after 47 h. At this stage, the deposits were not removed and the engine was switched over to 30% FCC+ 70% Reformate blend. The test was run for a further 20 hour duration. As the combustion chamber

32

Table - 8
Effect of Multifunctional Fuel Additive on Preignition

Fuel	Oil/Fuel ratio % vol	Test duration h	Exh. Port cleaned* No of times	S. Plug fouled Nos	Preign. at test duration,h	Nos.of Preign.
C	5	50	2	Nil	20, 47	2
F	5	50	2	Nil	9, 18	2

* Other than at occurrence of preignition when the engine power decreased to 90% of its initial value.

deposits were already built up in the sequence I, further engine operation for 20 hours was considered adequate to measure preignition propensity of the new fuel. During these additional 20 hours no preignition occurred confirming that even with high amount of combustion chamber deposits 30% FCC+70% Reformate had good resistance to preignition. When the engine was switched back to 100% FCC gasoline, preignition occurred again after 13.5 h of operation. The above test sequence was discontinued at this stage. This test sequence further demonstrated that with the fuels of high olefin content, preignition could occur in 2 - stroke engines when the conventional lubricants are used.

EFFECT OF MULTIFUNCTIONAL FUEL ADDITIVE - Multifunctional gasoline additives consisting of detergents and dispersants are being presently used with the objective to improve carburettor and induction system cleanliness (7). As mentioned earlier, one such additive formulation tended to reduce combustion chamber deposits and the deposit induced engine run on. The effect of one such commercial gasoline multifunctional additive was studied using the present test procedure. The preignition results with 100% FCC fuel treated with a multifunctional additive are compared in Table-8. This study showed no improvements in preignition behaviour with the use of a gasoline multifunctional additive Two preignition events were obtained and the exhaust port was cleaned two times, a result similar to that observed with the untreated FCC fuel. More detailed studies including the additive chemistry to evaluate the effectiveness of such additives are required with regard to preignition in 2-stroke engines.

CONCLUSIONS

From the present investigations on a small 2-stroke, gasoline engine the following findings are observed;

1- Increase in olefin content of gasoline shows increased tendency to preignition in air cooled crankcase scavenged 2-stroke gasoline engines.

2- Fuel olefin content of above 20% is seen to cause preignition during sustained operation of the engine at high load and speed conditions.

3- A typical multifunctional fuel additive formulation did not result in prevention of preignition with high olefinic gasoline.

ACKNOWLEDGMENT

The authors express their thanks to Dr. T S R Prasada Rao, Director, Indian Institute of Petroleum, Dehradun, for permission to publish this paper and to Shri T S Krishnamurthi, Executive Director, Centre for High Technology, New Delhi, for financial support to this program of work.

REFERENCES

1. A E Felt, J A Warren and C A Hall, "Rumble-A Deposit Effect at High Compression Ratios", Technical Papers of Ethyl Corporation, NewYork, 1958.

2. J C Guibet and A Duval, "New Aspect of Preignition in European Automotive Engines", SAE Paper 720114.

3. Holger Menrad, Manfred Harelhorst and Walter Erwig "Preignition and knock Behaviour of Alcohol Fuels" SAE Paper 821210.

4. V J Tomsic,"Surface Ignition Behaviour of Fuels"SAE Trans. Vol. 70, 1962.

5. E N Narayanan Kutty, "Automan India 1991", Association of Indian Automobile Manufacturers.

6. D W Gow and C E Trautman, "Lean Mixture Lubrication of Two Cycle Gasoline Engine", SAE Paper 660776.

7. L Pless, "Surface Ignition and Rumble in Engines", SAE Paper 650391, 1966.

8. J A Bert, J A Gething, T J Hansel, H K Newhall,R J Peyla and D A Voss, "A Gasoline Additive Concentrate Removes Combustion Chamber Deposits and Reduces Vehicle Octane Requirement", SAE Paper No.831709, 1983.

9. B K Gandhi, "Preignition in Two-Stroke Cycle Spark Ignited Engines," SAE Paper 371C, 1961.

10. C C Colyer and W L Sicker, "Two Cycle Engines Require Special Oils,"SAE Paper 317 A, 1961.

11. L O Bowman and R W Burchell, "Two Cycle Engine Prefer Ashless Detergents Oil" SAE Paper 65 V, 1959.

12. P Newman and D S Smith, "The Development and Testing of Gasoline Engine Two Stroke Oils for the European Market", SAE Paper 707A,1963.

13. Internal Combustion Engines and Air Pollution by Edward F Obert, Intext Educational Publishers, NewYork, 1973 Edition, PP 315-319.

14. CEC Test Procedure on Fiat 132C Spark Ignition Gasoline Engine for Preignition Tendencies of Engine Lubricants, CEC L-34-T-82, 1982 .

15. CEC Test Procedure on Piaggio Vespa 180 SS Engine for Evaluation of Two Stroke Engine Lubricants, CEC-L-21-T-77, 1977.

16. R L Mendiratta, M Gupta, G S Bisht and B P Pundir,"Report on Development of Bajaj Super Engine Test Method for Evaluation of Preignition Tendency of Two-Stroke Engine Lubricants", IIP Report No. EL 316.91, 1991.

APPENDIX - I

Preignition Test Conditions

Table - A 1

Test Phase	Duration h-min	Engine speed rpm	Load kW	SPG Temp. ^0C	Fuel Cons. kg/h	Remarks
Warm-up	0-10	1300$\pm$100 Idling	-	Record	Record	CO=3.5 to 4%
Conditioning	0-30	4000	3.2	Record	Record	
Test	49-30	4000	3.8$\pm$0.2	190$\pm$5	2.02$\pm$0.11	

The combustion chamber and spark plug seat temperatures and the engine torque are recorded continuously during the test.

(i) **Cylinder Head Temperature Stabilization**
After first 90 minutes of the test, the cylinder head temperature is taken to have stabilized.

(ii) **Indication of Preignition**
A rapid rise in cylinder head temperature above its stabilized temperature recorded after 90 minutes test duration, is an indication of major preignition. At this instant spark plug gasket (SPG) temperature also rises rapidly and torque decreases.

(iii) **Inspection, Decarbonization and Restart after Major Preignition**
   a) After major preignition it is necessary to stop the engine promptly. An automatic shut down is provided.
   b) Record the operating period upto the point of preignition.
   c) Remove the cylinder head, and inspect the cylinder bore. If there is no mechanical damage/seizure marks, decarbonise the combustion chamber and replace the cylinder head.
   d) Restart the engine and continue the test with new spark plug.

Note : If at any time during the test, the torque falls to 90% of its initial value, clean the exhaust port deposits, but if the power is not restored investigate and repair as appropriate.

932396

Application of Direct Air-Assisted Fuel Injection to a SI Cross-Scavenged Two-Stroke Engine

R. G. Kenny, R. J. Kee, C. E. Carson, and G. P. Blair
The Queen's University of Belfast

ABSTRACT

A 500 cc single cylinder two-stroke engine employing cross scavenging and direct air-assisted gasoline injection is described. Preliminary engine test results are presented for 3000 rpm full load and 1600 rpm part load operating conditions. The effects of fuel injection timing on full and part load brake specific fuel consumption and exhaust emissions are examined.

INTRODUCTION

This paper is concerned with the reduction of fuel consumption and exhaust emissions from two-stroke spark ignition engines. There is widespread interest in the application of direct in-cylinder fuel injection to improve the fuel consumption and exhaust emissions characteristics of two-stroke engines, particularly for automotive and outboard marine applications. The fundamental attractions of the two-stroke engine, compared to the four-stroke, are its potential for lower exhaust emissions and superior thermal efficiency, reduced weight, and greater compactness. For road vehicle applications the benefits in overall vehicle fuel economy and exhaust emissions from improved engine characteristics may be compounded by additional vehicle weight reductions and aerodynamic improvements, potentially derived from the more compact nature of the engine.

A number of experimental and theoretical investigations into direct injection two-stroke engines have been reported in the literature in recent years (1-24). It has been shown that, depending on operating conditions, the open cycle losses of fuel that are characteristic of the conventional mixture scavenged two-stroke engine may be reduced or even eliminated by directly injecting fuel into the cylinder. Ideally all the fuel should be injected in the period between exhaust port closing and ignition, thus eliminating the possibility of fuel short-circuiting. However, in practice, it is normally necessary at high loads to begin injecting fuel during the open cycle period because of the limited time available and the relatively low prevailing turbulence levels for mixture preparation .

When operating at full and medium loads with homogeneous combustion, the brake specific fuel consumption and exhaust hydrocarbon emissions may be greatly reduced by the application of direct fuel injection alone. Generally, at very light loads the high levels of charge dilution that exist under normal throttled operation prevent regular homogeneous combustion: a number of authors have identified that to achieve stability under these conditions it is desirable to operate in a direct injection stratified combustion mode. When operating in the stratified combustion mode, the injection system must deliver a finely atomised fuel spray to enable adequate mixture preparation within the cylinder before ignition. The use of both high pressure liquid injection (single fluid) systems and compressed air assisted (twin fluid) systems have been reported.

The majority of the studies in the literature have been concerned with loop scavenged engines. The aim of the present paper is to describe the initial fuel consumption and exhaust emissions test results achieved for direct in-cylinder air-assisted gasoline injection applied to a single cylinder research engine that employs the QUB cross scavenging system (25). The work described is part of a study undertaken to support the design and development of an in-line 3-cylinder direct injection two-stroke automotive research engine of 1500 cm^3 swept volume.

ENGINE DESIGN

The salient features of the single cylinder research

engine, designated as the QUB500RV, which was designed and built for the project are given in Table 1. A general arrangement drawing of the engine is shown in Fig. 1. The main details of the single cylinder engine were dictated by considerations relating to the 3-cylinder design. The engine is water cooled and has a square configuration with bore and stroke dimensions of 86 mm giving a swept cubic displacement of 500 cc. Induction is reed valve controlled and the scavenge and exhaust ports are piston controlled with port opening timings of 120° and 105° atdc respectively. The exhaust system is of the simple box silencer type. The crankcase and crankshaft assembly are adapted Kawasaki KX500 motocross engine parts. The engine employs a two-port variant of the QUB cross scavenging system, Fig. 2.

For the preliminary tests described here, the injector was installed in the cylinder head directly over the centre of the combustion cavity formed by the deflector piston as indicated in Fig. 1. The spark plug was located radially in the cylinder wall with its axis 17 mm below the TDC position. A trapped compression ratio of 6.9:1 and a squish clearance of 3.0 mm were used.

Engine Designation	QUB500RV
Type	DI Two-Stroke
No. of Cylinders	1
Cooling	Liquid
Bore	86 mm
Stroke	86 mm
Displacement	500 cm^3
Trapped Compressio ratio	6.9:1
Squish Clearance	3.0 mm
Induction System	Reed Valve
Exhaust Port Opens	105° atdc
Transfer Port Opens	120° atdc
Exhaust System	Box Silencer

Table 1.   Engine Specification

SCAVENGING SYSTEM - The QUB cross scavenging system uses a deflector piston, Fig. 3, which is heavier and creates a larger combustion chamber crevice volume than a typical piston for a loop scavenged engine. Nevertheless, cross flow scavenging has a number of features that merit its investigation for direct injection engines.

Cross flow scavenging enables very close cylinder spacing to be achieved without compromising scavenging efficiency or resorting to skewed cylinders, such as are commonly employed in multi-cylinder loop scavenged outboard marine engines. Close cylinder spacing not only facilitates a more compact design but, in addition, makes possible reductions in the lengths of the exhaust system's primary runners connecting the

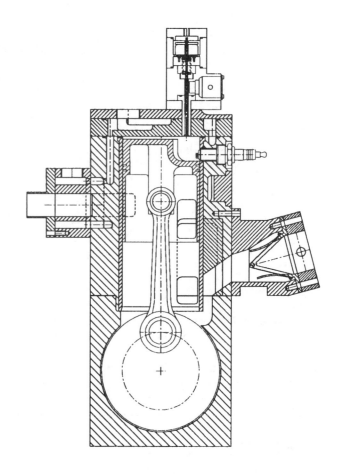

Fig. 1   QUB500RV single cylinder engine

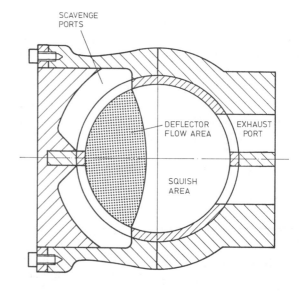

Fig. 2   QUB two-port cross scavenging layout

38

Fig. 3 QUB deflector piston

Fig. 4 Cylinder block

cylinders to the main exhaust pipe. In a 3-cylinder design it is usually not physically possible to reduce the lengths of the exhaust system primary runners to their theoretical optimal values, Fleck (26). Consequently, any reduction in the lengths of the primary runners will generally have a favourable impact on 3-cylinder engine exhaust tuning characteristics.

Good exhaust system tuning, i.e. the correct phasing of exhaust suction (expansion) and plugging (compression) pressure pulses at the open exhaust port, is necessary if high charging efficiencies and bmep levels are to be achieved over the full operating speed range. The phased arrival of an exhaust suction pulse at the open exhaust port during the scavenging period may boost the delivery ratio, and an exhaust plugging pulse correctly phased to arrive during the trapping period can raise the cylinder trapping pressure: the combined effect enhancing the overall charging efficiency. Also, because it is normally necessary at high loads to start fuel injection during the open cycle, with the attendant possibility for fuel short-circuiting, correctly phasing the inter cylinder exhaust plugging pulses may provide additional fuel consumption and hydrocarbon exhaust emission reductions.

The speed range over which exhaust tuning is effective may be extended with variable exhaust timing. In addition, there is some evidence to suggest that the internal fluid dynamic behaviour within the engine cylinder may also be altered by variable exhaust port timing to aid charge stratification and promote stratified combustion at light loads, Duret (13). By avoiding the use of skewed cylinder porting, the inclusion of an exhaust timing control device is eased for a multi-cylinder design.

Close cylinder spacing can be achieved on a loop scavenged engine without recourse to skewed cylinders if the main transfer passages are located immediately adjacent or underneath the exhaust porting. However such designs are likely to exhibit unnecessarily high levels of direct short circuiting with a detrimental influence on charging. Research (27) has shown that cross flow scavenging can be developed to achieve scavenging efficiency characteristics, as quantified by constant volume single-cycle testing (28), which are as good as or superior to the best loop scavenged designs.

CYLINDER BLOCK - The linered cylinder block, Fig. 4, employs an exhaust port with both vertical and horizontal bridges. The vertical bridge allows the use of a relatively wide exhaust port that has a cord width equivalent to 65 % of the cylinder bore. The horizontal bridge is incorporated in the design so that a late exhaust port opening of 135° atdc can be achieved for light load operation by throttling the upper pair of ports. It is believed that, if successful, a throttled port of this type would reduce the manufacturing complexity of the cylinder block and overcome possible durability concerns associated with the more normal exhaust timing control devices, such as those commonly employed on high performance and competition two-stroke engines.

AIR-ASSISTED INJECTION - The air-assisted fuel injector, Fig. 5, used for the tests described here was designed and manufactured in-house purely for research purposes. It employs a commercially available solenoid to control the operation of the main poppet valve, and a GM automotive low pressure liquid

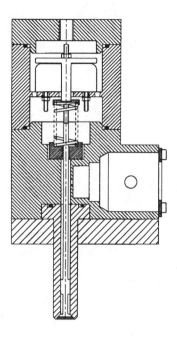

Fig. 5   Air- assisted fuel injector

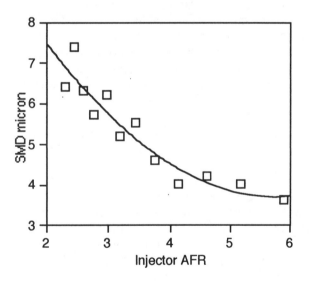

Fig. 6   Variation of Sauter Mean Diameter with injector AFR

fuel injector to meter fuel into the main injector body and provide a degree of pre-atomisation. A detailed description of a similar air-assisted injector and its operation is given in a previous publication (7). The diameter, lift, and seat included angle, of the air poppet valve were 6 mm, 0.35 mm and 90° respectively. The shape of the injector spray plume may be altered by a shroud on the nose of the injector: a 0.5 mm deep shroud was employed for the present work.

Droplet sizing tests were carried out on the injector using a Malvern Laser particle sizer. The injector was operated at a pulsing frequency of 1500 cycles per minute with air and fuel pressures of 550 and 650 kPa gauge respectively. Shell Sol-T was used as a fuel substitute. The air valve duration was fixed and the fuel duration varied. The injector spray was directed into an open space at atmospheric temperature and pressure. The droplet size measurements were made on the spray axis at 70 mm from the injector nozzle. Fig. 6 shows the variation of the measured Sauter mean diameter with injector air-to-fuel ratio (injector AFR). The results show that the droplet Sauter mean diameter remains low throughout the range of injector AFR values examined but rises at the lower end of the AFR range as might be expected from the work of Rizkalla and Lefebvre (29). A full study of the spray characteristics has not yet been undertaken. However, these results would seem to suggest that the fuel spray is well atomised.

Bench testing of the injector has also included valve dynamic response measurements. A 3 mm diameter nickle/cobalt magnet was bonded to the face of the poppet valve and a Hall Effect sensor was used to determine the response characteristics. The tests showed that the poppet valve opens approximately 2.5 ms after the signal is received from the driver circuit. This delay corresponds to 24° at 1600 rpm and 45° at 3000 rpm on the engine. In addition, the actual duration of the injector opening was shown to be approximately 50% greater than the nominal value. These delays are significant and thus the electronic or nominal SOA injection timings quoted should be adjusted accordingly to give actual mechanical timings for the start and end of the injection events.

Measurements of the solenoid current showed that the solenoid and the driver circuit were not correctly matched. It is believed that use of a solenoid with appropriate characteristics will significantly improve the speed of operation.

ENGINE TESTING

Steady state engine testing was carried out using a Froude eddy current dynamometer. Exhaust emissions were measured using an Oliver Multigas analyser which incorporated an FID instrument, calibrated to read in ppm propane, for hydrocarbons; NDIR instruments for CO and CO_2; a paramagnetic instrument for O_2; and a chemiluminescent instrument for NO_x.

WIDE OPEN THROTTLE TESTING - Wide open throttle (WOT) fuel loop tests were carried out at 3000 rpm with MBT ignition timing. The electronic triggering pulses which initiated the start of fuel metering (SOF) and the start of air injection (SOA) were timed to coincide. Thus the SOA effectively controlled the timing of fuel injection into the cylinder. The air valve open duration was fixed at 1.8 ms and the fuelling level was varied between the rich and lean limits by altering the value of the fuel pulse width. Fuel loops were repeated for different injection timings namely; SOA = 110°, 125°, and 140° atdc. A later injection timing of 155° atdc was attempted but stable engine operation could not be achieved. All duration and timing values, quoted for the injection process, are the nominal electronic values, unless otherwise stated. Fig. 7 shows bsfc; bsHC; bsCO; and $bsNO_x$ plotted against bmep. Also plotted are the values of delivered AFR, trapped AFR and fuel trapping efficiency calculated from exhaust gas analysis, as described by Douglas (30).

The peak bmep values obtained for the three SOA timings are all within about 0.1 bar of each other. There is a clear trend of reducing bsfc and bsHC with delaying SOA timing. This can be attributed to the higher fuel trapping efficiencies obtained with the later timings. It can be seen that with the nominal SOA timing of 110° atdc, the fuel trapping efficiencies lie for the most part at about 0.86 compared with about 0.94 for the later 140° timing. The poorer fuel trapping efficiencies exhibited by the early injection timing are reflected in lower values of delivered AFR for equal bmep levels.

The delivered AFR values vary in the range 16 to 26:1, depending on the load level and the SOA timing. In contrast, the variation with load of trapped AFR, which ranges from about 11 to 17:1, is very similar for each of the SOA timings considered. The bsCO - bmep, and the $bsNO_x$ - bmep, characteristics are essentially the same for all three SOA timings. The peak values of $bsNO_x$ occur at a trapped AFR of about 14.5:1.

The best overall results for brake specific fuel consumption and emissions are achieved with the 140 SOA timing. Taking into account the injected air, which by itself provided a delivery ratio of about 1%, the overall delivery ratio of the engine at this condition was 83 %. The peak bmep of 5.9 bar is obtained for a trapped AFR of 12.2:1 (19.2 delivered AFR), with corresponding values of bsfc and bsHC of 319 and 20.4 g/kWh respectively. The comparative values for bsCO and $bsNO_x$, both of which change rapidly at this condition, are about 140 to 190 g/kWh and 5.5 to 4.2 g/kWh respectively. The minimum values of bsfc, 293 g/kWh, and bsHC, 18.7 g/kWh, are obtained with the fuelling level reduced to give a bmep of 5.7 bar at a trapped AFR of about 14.5:1 (21.9:1 delivered AFR): at this condition the value of $bsNO_x$ peaks at 9.5 g/kWh and the bsCO is 41 g/kWh.

It is worth noting that at WOT, the AFR of the mixture delivered by the air-assisted injector was in the range 0.1 to 0.2. This is considerably richer than that considered in the droplet sizing experiments because of physical limitations in measuring droplet size in dense fuel sprays. The very low values of injector AFR at WOT are a consequence of the high fuel demand at this engine operating condition, and the limitations on injector airflow imposed by the injector design.

PART THROTTLE TESTING - Part throttle tests were carried out at 1600 rpm. It had originally been intended to explore the 1600 rpm/1.5 bar bmep operating point corresponding to a typical automotive low speed light load mapping point. However, with the engine in its initial build as described here, stable stratified operation could not be achieved at this speed/load condition. Accordingly, the fuelling was varied for a number of fixed part throttle positions to understand better the engine's behaviour and provide baseline information for subsequent development work.

Fig. 8 shows the results of tests carried out at a nominal delivery ratio of 0.4 with a constant SOA injection timing of 250° atdc. The tests were repeated for MBT and 35° btdc ignition timings, and for two exhaust opening timings; namely, 105° and 135° atdc. The 35° ignition timing typically represents an ignition advance of 10° to 15° from the MBT condition. The later exhaust port opening was achieved by blocking the upper pair of exhaust ports with machined aluminium inserts.

With MBT ignition timing and EO = 105° atdc, a minimum bsfc of 305 g/kWh is obtained at a bmep of 2.7 bar . The advanced ignition timing of 35° btdc enables stable operation to be achieved at a lower bmep level, but causes a deterioration in the minimum bsfc to about 365 g/kWh at around 2 bar bmep. The effect of the reduced exhaust duration is to marginally improve the bsfc: for example, with MBT ignition, the minimum bsfc is reduced from 305 to 293 g/kWh.

The delivered AFR varies from 15 to 40:1, whereas the trapped AFR varies over the range 10 to 25:1. It is noted that the trapped AFR is about 15:1 at the minimum bsfc points.

When operating with the ignition timing advanced, the trapped AFR - bmep characteristics show a richer AFR requirement for a given load: this is a reflection of the increase in fuelling needed to compensate for the greater negative cylinder work compared with the MBT case. The richer trapped mixture promotes more stable combustion at the lighter loads. This has consequences for the bsHC, bsCO and $bsNO_x$ characteristics.

It can be seen for these tests that the bsHC emissions are not greatly influenced by the change in EO timing from 105° to 135° atdc. The minimum bsHC for MBT ignition, 9 g/kWh, was obtained at a slightly higher load level (and richer trapped AFR) than the minimum bsfc. With the advanced ignition timing, the bsHC - bmep characteristic curves are shifted towards the lower loads, presumably because of the improved

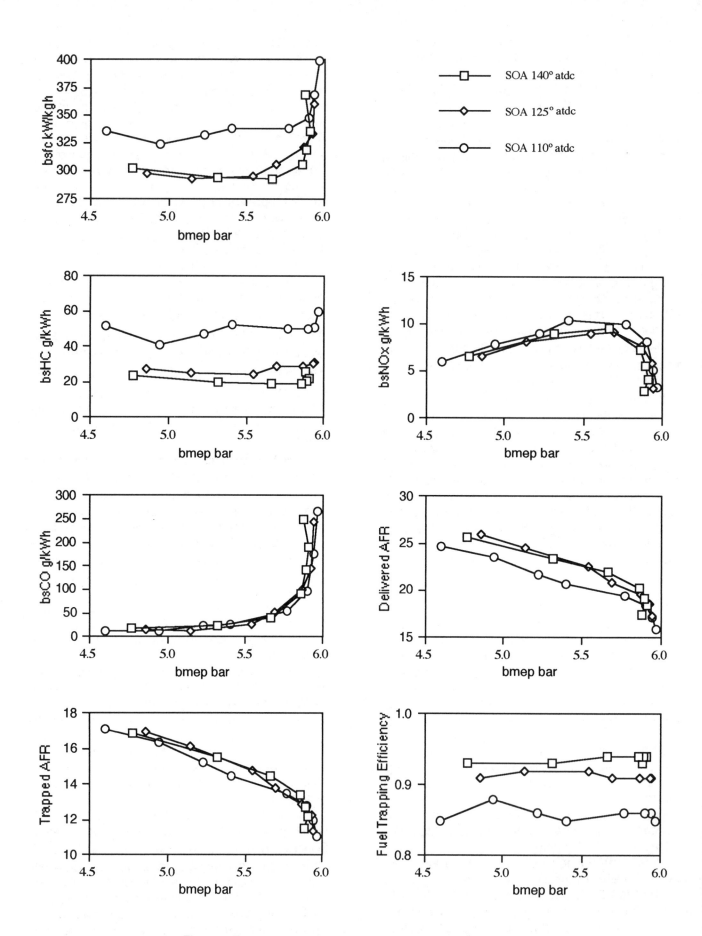

Fig. 7   Engine Performance characteristics for WOT at 3000 rpm

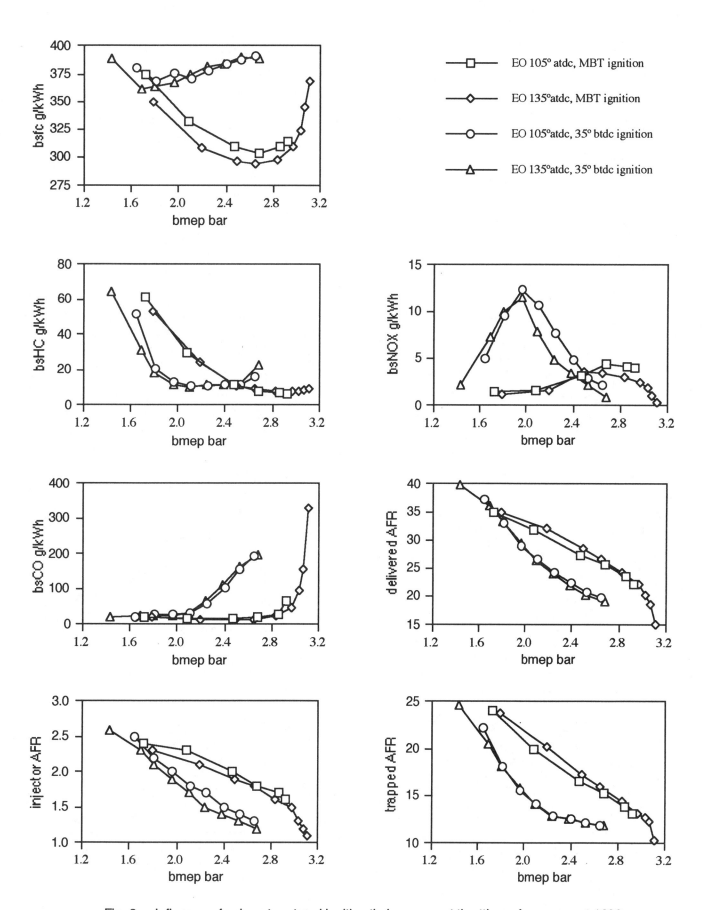

Fig. 8   Influence of exhaust port and ignition timings on part throttle performance at 1600 rpm,
DR = 0.4, and SOA = 250° atdc

43

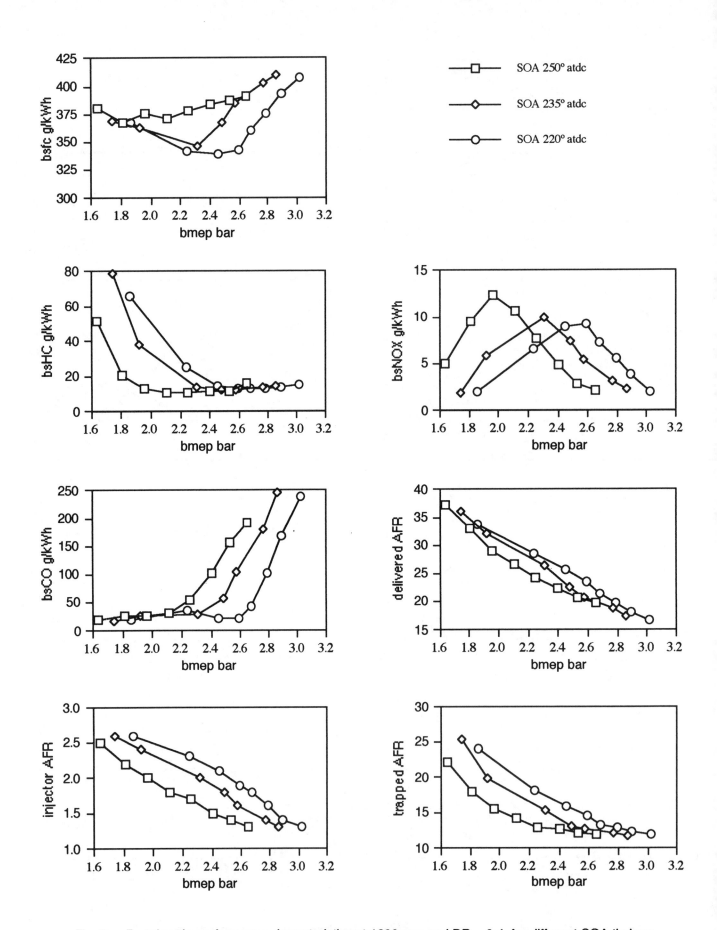

Fig. 9   Part throttle performance characteristics at 1600 rpm and DR = 0.4 for different SOA timings

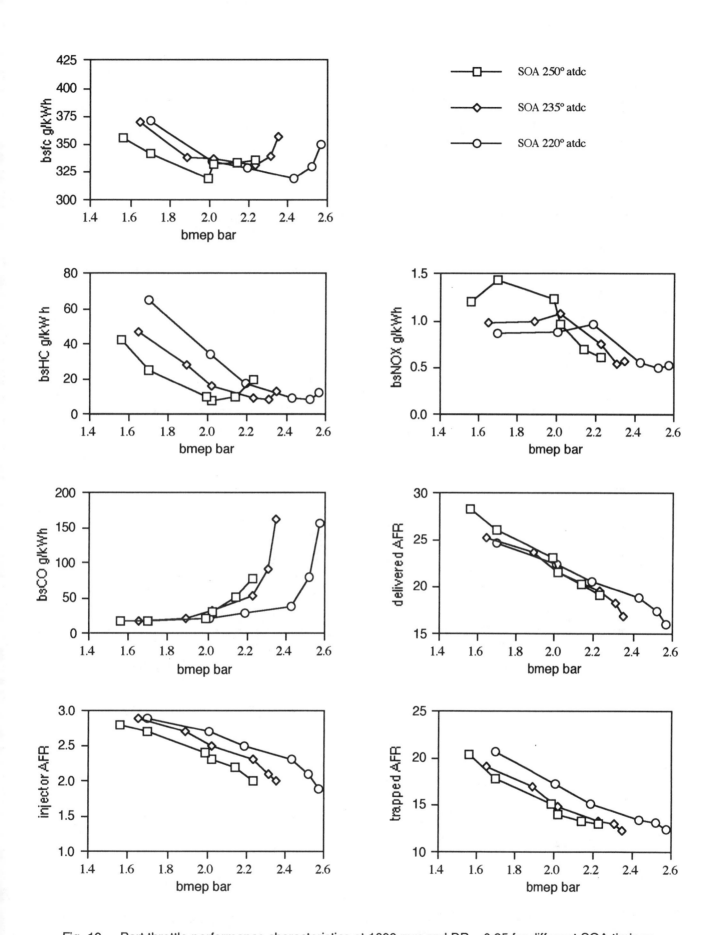

Fig. 10　Part throttle performance characteristics at 1600 rpm and DR = 0.25 for different SOA timings

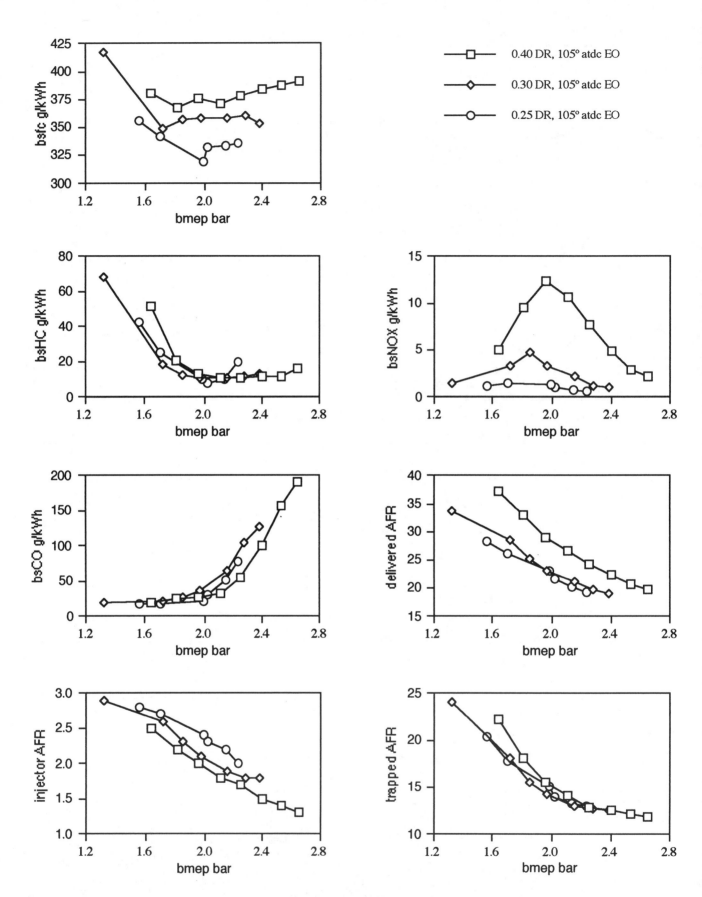

Fig. 11   Performance characteristics for different values of DR with SOA = 250° atdc, EO = 105° atdc, and 35° btdc ignition timing

combustion stability.

The variation of bsCO with bmep is broadly similar for the two exhaust timings, but differs for the two ignition timings. As happens for the bsHC emissions, the bsCO - bmep characteristic curves are shifted to the left (i.e. towards the lower bmep levels) for the 35° btdc ignition timing. The bsCO is seen to level off at the lower loads, where the trapped AFR rises above the stoichiometric value, i.e. below 2 and 2.7 bar for the 35° btdc and MBT ignition timings respectively.

The bsNO$_x$ emissions peak at over 10 g/kWh for the 35° btdc ignition, whereas with MBT ignition the peak values are below 5 g/kWh. The effect of the reduced exhaust open duration is to reduce the bsNO$_x$, emissions. It is noted that at peak bsNO$_x$ the corresponding trapped AFRs are close to stoichiometry.

Figures 9 and 10 show the effect of SOA injection timing for constant delivery ratios of 0.4 and 0.25 respectively. An ignition timing of 35° btdc and the standard 105° EO timing were employed for these tests. In both cases, the general effect of retarding the SOA timing is to shift the characteristic curves to the left and to improve the light load stability and bsHC. There is also a general trend with retarding SOA timing of increasing minimum bsfc, decreasing minimum bsHC, and increasing peak bsNO$_x$. However, the differences are relatively small.

At the 0.4 delivery ratio, Fig. 9, the minimum bsfc, and peak bsNO$_x$ values occur at approximately 2.5, 2.3 and 2 bar for the 220°, 235° and 250 SOA timings respectively: in each case the corresponding trapped AFR value is about 15:1. The minimum bsHC for each of the SOA timings are recorded at slightly higher loads and richer values of trapped AFR. The overall minimum bsfc, 340 g/kWh, is produced by the earliest of the three injection timings, i.e. SOA = 220° atdc. It is notable that the later 250° SOA timing allows lower bsHC values to be achieved at the lighter load levels, for example 13 g/kWh at 2 bar: however, peak bsNO$_x$ at 12 g/kWh was greatest for this SOA timing.

A similar pattern is repeated in Fig. 10 for the 0.25 delivery ratio; namely lowest bsfc for the early SOA timing, lower bsHC emissions at the lower loads achieved with the latest SOA timing, and higher peak bsNO$_x$ emissions at the latest SOA timing. Notably, the bsNO$_x$ emissions are reduced by an order of magnitude from those for the 0.4 delivery ratio tests, Fig. 9.

The influence of delivery ratio is further illustrated in Fig. 11, where results are plotted for 0.4, 0.3 and 0.25 delivery ratios. These tests were all carried out with EO = 105° atdc, an ignition timing of 35° btdc, and SOA = 250° atdc. The bsfc can be seen to improve with reducing delivery ratio, with a minimum of 330 g/kWh being obtained at a bmep of about 2 bar with DR = 0.25.

At the 2 bar point, the bsHC emissions and the bsCO emissions are very similar for all three delivery ratios. The bsNO$_x$ emissions are lowest for the 0.25 delivery ratio, with a maximum value of 1.5 g/kWh.

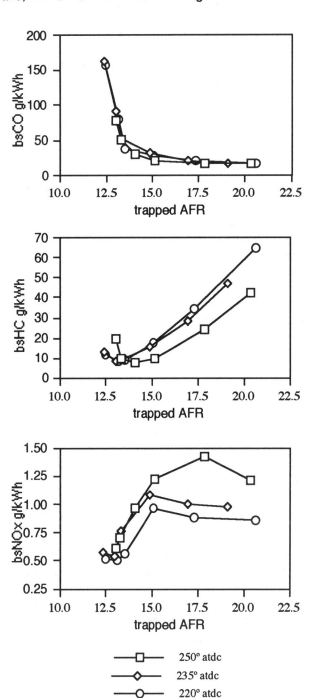

Fig. 12   Influence of trapped AFR on brake specific emissions for different SOA timings with DR = 0.25, EO = 105° atdc, and 35° btdc ignition timing

DISCUSSION

The 3000 rpm wide open throttle results obtained confirm that good thermal efficiencies can be achieved with direct injection two-stroke engines. Provided durability is acceptable, it would be preferable to operate on the lean side of the peak bmep to reduce both bsfc and bsHC. The present single cylinder engine produces a respectable 5.7 bar bmep when operating with a near stoichiometric trapped AFR. It should be emphasised that this bmep is obtained without the benefit of significant favourable exhaust tuning. A three-cylinder engine with good exhaust tuning would be expected to produce about 7.5 bar bmep at a similar speed.

The bsHC characteristics show a 5 to 10 times reduction compared with the equivalent homogeneous mixture scavenged results (31). Nevertheless, the minimum value of bsHC, 19 g/kWh, is disappointing in comparison with published data for four stroke engines. The inherently high crevice volume associated with the QUB deflector piston design may be a contributory factor to these relatively high hydrocarbon emissions. However, the minimum bsHC of a previous engine (7) of a broadly similar design, but of 270 cc capacity, was approximately half the bsHC of the present engine, when tested under the same conditions apart from the injector AFR. The injector AFR of the previous engine was significantly higher due to the smaller fuel demand resulting from the lower cylinder capacity. The detailed optimisation of the fuel injector spray characteristics, injector position and orientation, and basic combustion chamber geometry, has yet to be attempted and should provide scope for abatement of the bsHC emissions at WOT and part throttle operation.

Table 2 shows the estimated mechanical SOA timings for the 3000 rpm WOT tests based on the bench testing. All of the estimated SOA injection timings occur well before exhaust port closure, i.e. 105° btdc. Accordingly, further improvements in WOT bsHC (and bsfc) should be possible with exhaust tuning because of the essentially open cycle nature of the fuel injection process at WOT.

The WOT bsNO$_x$ emissions, 9 g/kWh at the near stoichiometric trapped AFR operating condition, are probably about one quarter of the equivalent untreated four-stroke full load values and are in line with other published direct injection two-stroke engine data (23).

At part throttle, advancing the ignition timing beyond MBT improves stability and lowers the hydrocarbon emissions for the lighter loads. The effect of reducing the exhaust duration is to improve bsfc and bsNO$_x$. As can be seen from Table 3 the 250° atdc SOA timing employed for these tests corresponds to an estimated mechanical SOA timing of 274° atdc (or 86° btdc), which is after both the standard and reduced

duration exhaust ports close. Hence all the delivered fuel is trapped under these conditions. Consequently the improvements in bsfc and bsNO$_x$ may be attributed to the increased expansion stroke and greater exhaust gas retention.

Nominal SOA ° atdc	Mechanical SOA ° atdc (° btdc)
140	185 (175)
125	170 (190)
110	155 (205)

Table 2.   Estimated mechanical SOA at 3000 rpm WOT

Nominal SOA ° atdc	Mechanical SOA ° atdc (° btdc)
250	274 (86)
235	259 (101)
200	244 (116)

Table 3.   Estimated mechanical SOA at 1600 rpm part throttle

The SOA injection timing surveys carried out at 1600 rpm for DR = 0.4 and 0.25 show that the light load stability and bsHC can be improved by retarding the the SOA timing, in this case to 250° atdc. Fig. 12 shows the brake specific emissions data for the DR = 0.25 case plotted against trapped AFR. The bsHC values are lower for the late injection timing at the lean values of trapped AFR. Also, the peak bsNO$_x$ occurs at a lean trapped AFR of about 17.5:1. These results suggest that some degree of stratified combustion is being achieved at this low delivery ratio with the late injection timing. It is believed that increases in the spray droplet sizes due to decreasing injector AFR at late injection timings, and the inadequent dynamic response characteristics of the present injection system, are important factors preventing stable stratified combustion operation at later injection timings.

Overall, the best part load results are achieved with a delivery ratio of 0.25. At this low delivery ratio, the high levels of retained burned gas enable particularly low levels of NO$_x$ to be achieved. It is, of course, particularly important to attain low untreated bsNO$_x$ emissions because of the difficulty in catalytically reducing NO$_x$ with the excess oxygen normally prevailing in two-stroke engine exhaust gases.

CONCLUSIONS

A single cylinder two-stroke engine employing QUB cross scavenging and direct air-assisted fuel injection has been built. The results of initial WOT and part throttle testing have shown:

1.   A peak bmep of 5.9 bar was obtained at WOT using a simple box silencer exhaust system. A slightly lower bmep of 5.7 bar was obtained with a stoichiometric trapped AFR.

2.   Good thermal efficiencies and low brake specific NO_x emissions can be achieved at WOT operation using a stoichiometric trapped AFR.

3.   At WOT the optimum timing for the start of fuel injection was found to be in the open cycle period.

4.   Stable operation could not be achieved at the desired 1600 rpm / 1.5 bar light load mapping point with the initial engine build specification described here.

5.   Advancing the ignition timing beyond MBT at part throttle and light loads was found to improve stability and reduce the brake specific hydrocarbons but deteriorated the bsfc.

6.   Retarding the exhaust opening from 105° to 135° atdc improved the bsfc and $bsNO_x$ emissions at part load.

7.   Retarding the start of the fuel injection process to SOA = 250° atdc improved the light load stability and bsHC.

8.   The high levels of retained burned gas associated with operation at a low delivery ratio of 0.25 enabled particularly low levels of $bsNO_x$ to be achieved at light loads.

ACKNOWLEDGEMENTS

The authors would like to thank The Queen's University of Belfast and the Department of Mechanical and Manufacturing Engineering for the provision of laboratory facilities. Thanks are also due to the Department of Education for Northern Ireland for their financial support.

REFERENCES

1.   Huei-Huay Huang, Ming-Horng Jeng, Nien-tzu Chang, Yue-Yin Peng, James H. Wang and Wei-Li Chiang, "Improvement of Exhaust Emissions of a Two-Stroke Engine by Direct Injection System", SAE Paper No. 930497, 1993.

2.   Paul Conlon and Robert Fleck, "An Experimental Investigation to Optimise the Performance of a Supercharged Two-Stroke Engine", SAE Paper No. 930982, 1993.

3.   R. Diwakar, Todd D. Fansler, Donald T. French, Jaal B. Ghandhi, Cameron J. Dasch, David H. Heffelfinger, "Liquid and Vapour Distributions from an Air-Assist Injector - An Experimental and Computational Study", SAE Paper No. 920422, 1992.

4.   Twang-Wei Kuo and R. D. Reitz, "Three-Dimensional Computations of Combustion in Premixed-Charge and Direct-Injected Two-Stroke Engines", SAE Paper No.920425, 1992.

5.   Duane Abata and Kurt Wellenkotter, "Characterization of Ignition and Parametric Study of a Two-Stroke-Cycle Direct-Injected Gasoline Engine", SAE Paper No. 920422, 1992.

6.   K. C. Schlunke, S. R. Ahern, S. R. Leighton, and M. R. Kitson, "Fuel Consumption and Emissions in Small Displacement Two-Stroke Cycle Engines", Fourth Graz Two-Wheeler Symposium, VKM-Thd, Graz University of Technology, Graz, 8-9 April 1991, Vol 60b, p. 142.

7.   G. P. Blair, R. J. Kee, C. E. Carson, and R. Douglas, "The Reduction of Emissions and Fuel Consumption by Direct Air-Assisted Fuel Injection into a Two-Stroke Engine", Graz University of Technology, Graz, 8-9 April 1991, Vol. 60b, p. 114.

8.   Huei-Huay Hang, Yu-Yin Peng, Ming-Hong Jeng, and James H. Wang, "Study of Small Two-Stroke Engine with Low-Pressure Air-Assisted Direct-Injection System", SAE Paper No.912350, 1991.

9.   Gaëtan Monnier and Pierre Duret, "IAPAC Compressed Air Assisted Fuel Injection for High Efficiency Low Emissions Marine Outboard Two-Stroke Engines", SAE Paper No. 911849, 1991.

10.   Michael M. Schechter and Michael B. Levin, " Air-Forced Fuel Injection System for 2-Stroke D.I. Gasoline Engine", SAE Paper No. 910664, 1991.

11.   Michael M. Schechter, Eugene H. Jary, and Michael B. Levin, "High Speed Fuel Injection System for Two-Stroke D.I. Gasoline Engine", SAE Paper No. 910666, 1991.

12.   J. W. Adams, R. A. Stein, G. F. Leydorf, Jr; and M. J. Schrader, "Initial Evaluation of a Spill Valve Concept for Two-Stroke Cycle Light load Operation", SAE Paper No. 901663, 1990.

13.   P. Duret and J-F Moreau, "Reduction of Pollutant

emissions of the IAPAC Two-Stroke Engine with Compressed-Air Assisted Fuel Injection", SAE Paper No. 900801, 1990.

14. G. E. Hundleby, "Development of a Poppet-valved Two-Stroke Engine - the Flagship Concept", SAE Paper No. 900802, 1990.

15. M. Nuti, "A Variable Timing Electronically Controlled High Pressure Injection System for 2-Stroke S.I. Engines", SAE Paper No.900799, 1990.

16. K. Landfahrer, D. Plohberger, H. Alten, and L. A. Mikulic, "Thermodynamic Analysis and Optimization of Two-Stroke Gasoline Engines", SAE Paper No. 890415, 1989.

17. K. Schlunke, "Der Orbital Verbrennungsprozess des Zweitaktmotors", Tenth International Symposium on Engines, Vienna, April 1989, p.63.

18 P. Duret, A. Ecomard, and M. A. Audinet, "A New Two-Stroke Engine with Compressed-Air Assisted Fuel Injection for High Efficiency Low Emission Applications", SAE Paper No. 880176, 1988.

19. T. Sato and M. Nakayama, "Gasoline Direct Injection of a Loop-Scavenged Two-Stroke Cycle Engine", SAE Paper No. 871690, 1987.

20. V. Kuentscher, "Application of Charge Stratification, Lean Burn Combustion Systems and Anti-Knock Control Devices in Small Two-Stroke Cycle Gasoline Engines", SAE Paper No. 860171, 1986.

21. Marco Nuti, "Direct Fuel Injection: An Opportunity for Two-Stroke SI Engines in Road Vehicle Use", SAE Paper No. 860170, 1986.

22. N. John Beck, W. P. Johnson, R. L. Barkhimer, and S. H. Patterson, "Electronic Fuel Injection for Two-Stroke Cycle Gasoline Engines", SAE Paper No. 861242, 1986.

23. Brian Cumming, "Two-Stroke Engines -Opportunities and Challenges", Aachen Colloquium on Automobile and Engine Technology, Institut fur Kraftfrahrwesen, Lehstruhl fur Angewandte Thermodynamic, Aachen. 1991, pp11.

24. P. Duret, S. Venturi, and Ch. Carey, "The IAPAC Fluid Dynamically controlled automotive two-stroke combustion process", VDI- IVK Symposium, A Comparison of automobile engines, Dresden, June, 1993.

25. G. P. Blair, R. A. R. Houston, R. K. McMullan and

S. J. Williamson, "A New Piston Design for a Cross-Scavenged Two-Stroke Cycle Engine with Improved Scavenging and Combustion Characteristics", SAE Paper No. 841096, 1984.

26. R. Fleck, "Three Cylinder, Naturally Aspirated, Automotive Two-Stroke Engines a Potential Performance Evaluation", SAE Paper No. 901667, 1990.

27. G. P. Blair, R. G. Kenny, J. G. Smyth, M. E. G. Sweeney and G. B. Swann, "An Experimental Comparison of Loop and Cross Scavenging of the Two-Stroke Cycle Engine", SAE Paper No. 861240, 1986.

28. M. E. G. Sweeney, R. G. Kenny, G. B. G. Swann, and G. P. Blair, "Single Cycle Gas Testing Method for Two-Stroke Engine Scavenging", SAE Paper No. 850178, 1985.

29. A. A. Rizkalla, and A. H. Lefebvre, "Influence of liquid properties an airblast atomization", Trans ASME, J. Fluids Engng, 97, 316-320, 1975.

30. Roy Douglas, "AFR and Emissions Calculations for Two-Stroke Cycle Engines", SAE Paper No. 901599, 1990.

31. G. P. Blair, R. J. Kee, R. G. Kenny, and C. E. Carson, "Design and Development Techniques Applied to a Two-Stroke Cycle Engine", VDI- IVK Symposium, A Comparison of automobile engines, Dresden, June, 1993.

932397

Factors Affecting Catalyst Efficiency a Theoretical and Investigative Treatise

B. P. Carberry and R. Douglas
The Queen's University of Belfast

ABSTRACT

This paper details the investigation of the properties of inlet gases and shows how they affect the flow patterns immediately in front of the catalyst and the subsequent loss of efficiency. A thorough analysis of the flow distribution at the inlet of the catalyst enabled the effective catalyst diameter to be calculated. Subsequent calculations were then carried out to determine the loss of catalyst function through flow maldistribution.

Experimental work involved flowing engine proportioned amounts of air through canisters of a fixed geometric profile containing a catalyst. Inlet cones of angles 10°, 15° and 45° were flowed to estimate the effect of the cone design on the velocity distributions at the face of the catalyst. Simple geometric profiles were investigated to allow a thorough understanding of the mechanism of flow to be comprehended and its affect on catalyst conversion to be analysed. A new ratio has been defined to assess the ability of the cones to develop the flow sufficiently enough to considerably increase the catalyst function.

Flow distribution has various overtones in the actual durability of the catalyst substrate namely:

1. Partial catalyst usage of theoretical flow area.
2. During intensive periods of hydrocarbon concentration in exhaust gas, after blowdown, inexact canister design will contribute to:

a. Large exotherms due to increased localised concentration of fuel enriched gas, leading to excessive degradation and substrate sintering.
b. With the concentrated flow profile the catalyst will experience excessive space velocities that can exceed their designed values by as much as 23%. This leads to the presence of unconverted hydrocarbons after the catalyst known as break through.

This information when interfaced with a mathematical model of the catalyst causes a closer correlation between previously published measured and predicted results.

INTRODUCTION

Catalysis has provided one of the most realistic methods of decreasing the levels of exhaust gas species legislated against by Economically Developed Countries. With the adoption of the catalytic converter in the automotive sector in 1975, much work has been done in perfecting the unit to increase its specific efficiency. However, new problems face the catalytic converter especially with the fuel enriched exhaust gas of the small capacity carbureted two-stroke-cycle engine. Another factor which continues to inflict serious damage to the catalyst is the threat of misfire. With its subsequent fuel enriching of the exhaust gas, which will impart a thermal shock to the substrate once ignited.

These two life abating factors of the catalyst are somewhat related, the significant difference being that the short circuited fuel in the carbureted two-stroke cycle engine is continuously present in the exhaust gas mixture. This is especially true of the exhaust gas produced by small capacity engines.

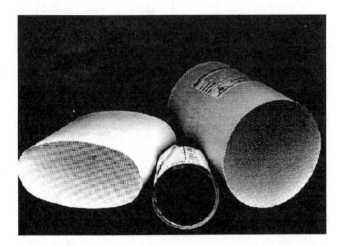

Fig 1. Ceramic and Metallic Catalyst Substrates.

The consequence of fuel enrichment in the exhaust gas will remain one of the main contributors to the loss of catalyst function. This has been primarily due to the loss of surface area due to excessive sintering from intensified local reaction phenomena during the hydrocarbons (HC) oxidation. It is quite apparent that the presence of high concentrations of fuel enriched exhaust gas is a condition that the catalyst must co-exist with especially from the small capacity carbureted two-stroke-engine. Currently a debate is in progress over the preference of which of the two types of carrier should be used, see Fig (1); namely

1. Ceramic.
2. Metallic.

This research paper does not tend to take issue with either of these schools of thought but rather what can be done to increase the:

1. The conversion efficiency of the catalyst
2. The durability of the catalyst employed.

Increasing Efficiency

In general the efficiency of the catalyst is dependent upon two parameters, namely the physical formulation and the nature of the flowing gas and its associated properties. Catalyst formulation is also a combination of two distinct areas namely washcoat technology and precious metal coverage. Therefore in specifying the catalyst to be used for a particular application, knowledge is required not only of the interaction of the gases with the precious metal, but also of how the precious metal interacts with the catalyst washcoat.

The performance of catalyst formulations has increased significantly over the past twenty years as surmised by Church[1] and Gulati[2], due to the advent of new additives which can be used to enhance the kinetic reaction rates of the catalyst. With the addition of ceria oxide as discussed by Oh[3] to the formulation of the washcoat the catalyst may now be exposed to a redox fluctuating exhaust gas stream. This additive still permits the oxidation of the HC present due to its oxygen storage capability.

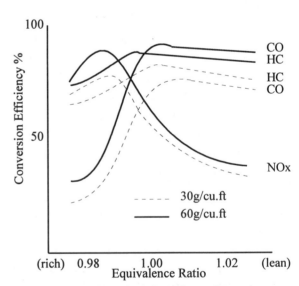

Fig 2. Conversion as a function of Precious Metal Loading.

With the addition of the above mentioned additive along with the precious metals to the washcoat, it would be expected that any exhaust gas regardless of its pollutant concentration could be decreased. However what must be taken into account is the interaction of these additives on the performance of the overall catalyst. Initial design criteria used by engineers to remove CO and HC from the exhausted smoke stacks at chemical plants, took into account the presence of a precious metal and its capability of oxidising a known exhaust gas constituent. Along with the flow rate of the gas the conversion efficiency could then be predicted. This early attempt to predict the conversion efficiency provided a close correlation because of the steady flow nature of the exhaust gas and the consistancy of its formulation. However when applied to an unsteady exhaust gas it provided to be a poor predictive tool because it failed to account for the highly competitive nature of the gases to be oxidised on the precious metal. Another factor which influences the conversion efficiency is the effect of precious metal loading on the washcoat surface, Fig (2). By increasing the density (loading) of precious metal on the washcoat, oxidation can occur more easily due to the increased capacity for adsorption of the gas, thus substatantially lessening the competition for the precious metal "active" sites.

Increasing Durability

Altering the washcoat formulation and precious metal loading to increase the conversion efficiency of the catalytic converter, will however increase only the short term efficiency, whilst increasing the overall component cost. An alternative method of promoting the efficiency and durability of the catalyst would be to investigate how well the catalyst is being used.

Typically a catalyst is sized using the following equation.

$$SP.Vel = \frac{B}{CV} \qquad EQ(1)$$

The volume CV assumes that there is no maldistribution of gas across the front of the catalyst, therefore the average packet of gas will have a mean residence time within the catalyst as defined in equation EQ(2).

$$Res_{Time} = \overline{\frac{1}{(SP.Vel)}} \qquad EQ(2)$$

If the flow rate in one particular section of the catalyst appears to be lower than the average flow rate, a longer residence time will result for the exhaust gas to be in intimate contact with the catalyst. Ultimately, this increases the probability that a higher conversion efficiency will occur. Although this may seem initially advantageous, a localised increase in conversion efficiency may be a trade off for a higher concentration of exhaust gas being channelled elsewhere in the catalyst, thus promoting what is known as breakthrough. Breakthrough describes the action of a gas when its flow rate is too great to allow catalytic conversion to occur.

Therefore this treatise will primarily investigate and quantify the maldistribution of exhaust gas flow for small capacity utility engines. Where a slight increase in uniform catalyst usage will produce a greater increase in conversion efficiency. Secondly the patterns of flow will be analysed to ascertain the uniformity of flow. The recorded data is then employed to assess the flow efficiency of the catalyst housing and then applied to correct the predictive capability of the catalyst model by Carberry, Douglas[4] presented previously to this society.

PART I

Investigating Flow Distribution

Catalyst Sizing. The initial sizing of the catalyst for this investigation was based on its application to a small capacity 2-stroke cycle engine typically of capacity 50 cc. Various modifiers may be used to tailor the catalyst size, however in this investigation only a space velocity of 120000 will be used.

Application Details
Engine Capacity (cc)   50
Max. Engine Speed (rpm)   9000
Max. Flow Rate (m^3/hr)   27

Catalyst Details

Space Velocity (hr^{-1})	120000
Diameter (mm)	51
Length (mm)	50
No. Cells	400

Flow Rig Development. To permit full comprehension of the effects of flow maldistribution, pulsating exhaust gas was not used but rather a steady flow of air produced by a variable flow centrifugal blower. This blower was in turn connected to a hot wire air flow metering device. This ensured that the catalyst received the correct volumetric flow of air for each set of tests, Fig(3).

The catalyst was firmly held in an fixture which permitted various cone geometries to be inter-changed, Fig (4). In this present series of tests the catalyst substrate was used to provide the perfect pitot tube assembly for measuring the flow, Fig (5). The flow normal to the front face of the catalyst was measured by sealing a small hypo-dermic needle into each cell of the substrate, the cell then acting as the collecting capillary. In addition to this the static wall pressure was also measured.

TEST PROCEDURES

To ensure that the presence of the fine capillaries at the rear of the catalyst had not disturbed the normal flow by inducing a back pressure in the canister the following procedure was adopted. The unit was flow tested with one capillary in place and then several on a radial profile to ascertain any change of reading. The measurement of the total pressure coupled with the static pressure ensured that the velocity profile adjacent to the catalyst substrate face could be defined. The investigation was carried out into the effects of different inlet geometry on the distribution of flow on the substrate face. During a test the substrate was positioned 2 mm from the end of the diffuser. The above assembly was then supplied with air at steady flow rates equivalent to certain defined speed settings. Assuming stoichiometric combustion there will be a mass increase of flow due to the addition of

fuel during the combustion stage. This approximates to a factored increase of 1.26. The rates of flow are also subject to the delivery ratio of the engine being simulated. The test points to be investigated are listed in Table 1.

Test	RPM	m³/hr
1	3000	8.5
2	6000	16.1
3	9000	27.1

Table 1. Experimental Flow Rates Used.

The inlet configurations that were tested are indicated in Table 3 along with the relative dimensions of each section, Fig(8) and Fig(4) show schematics of the diffusers used.

Geom Tested	Cone Angle	Inlet Length	Diffuser Length	Body Length
1	10	35	60	35
2	15	45	45	50
3	45	70	12	55

Table 2. Description of Diffuser Geometry's.

Each geometry was flow tested and if fluctuation occurred during measurement the average of the two extreme values were taken.

THEORY

To ascertain the flow distribution, measurements were taken using the above methodology. As stated previously, the pressure measured in front of the catalyst at the entrance to the substrate cell is the summation of dynamic and static head, the total pressure EQ(3).

$$P_{TOTAL} = P_{STATIC} + 1/2 \, \rho v^2 \qquad EQ(3)$$

By rearranging this equation the incident velocity of flow can be determined by EQ(4)

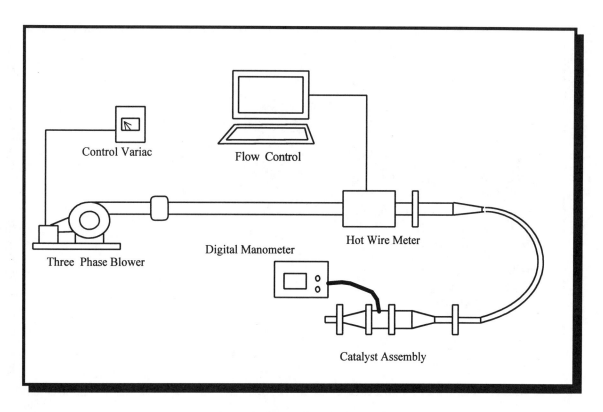

Fig. 3 Schematic of Flow Rig.

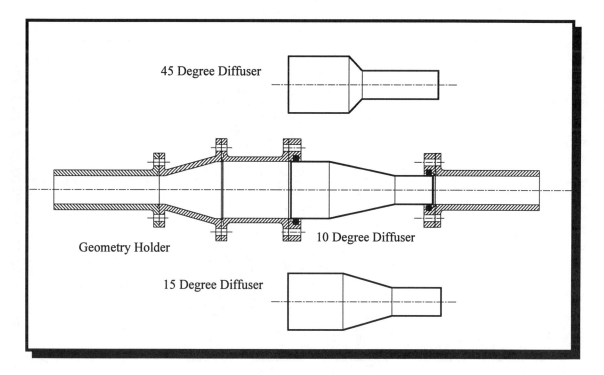

Fig.4 Inlet Geometry Holder

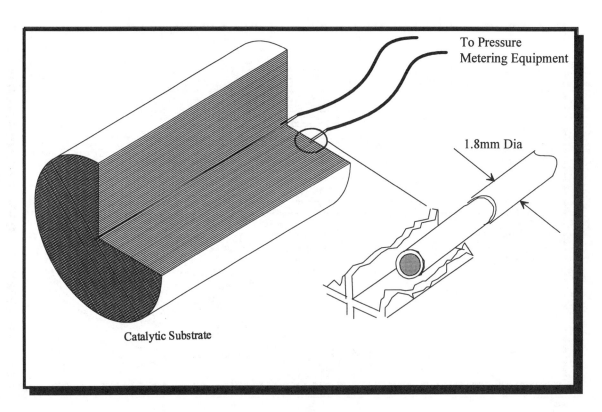

Fig 5. Catalytic Substrate Showing Pitot Positions.

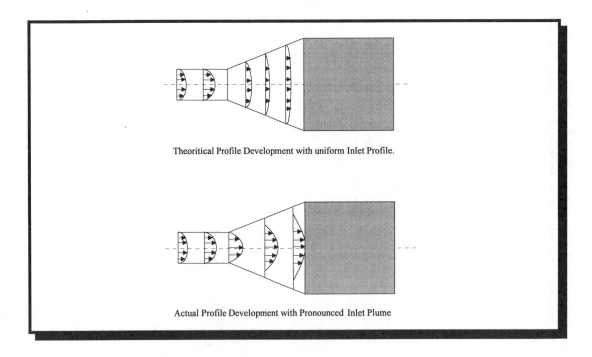

Fig 6. A Comparison of Flow Developments.

The resulting flow which is incident to the face of the catalyst will consist of a velocity profile which will differ from the assumed uniform velocity when using EQ(1) see Fig (6).

$$v = \sqrt{\frac{2 * (P_{TOTAL} - P_{STATIC})}{\rho}} \qquad \text{EQ(4)}$$

Therefore it is proposed to establish a coefficient of flow that will permit the flow from the three different inlet geometries to be assessed, quantified and compared.

As stated the flow incident to the substrate is a function of the radial displacement from the axis of the substrate. The inlet geometries which are dealt with provide flow symmetry about the axis of the inlet. Numerical determination of the degree of flow distribution exhibited on a frontal catalyst area has produced many but varied indices. These have ranged from those that have employed a statistical methodology, such as Weltens[5], Baxendale[6] and Lai[7], to those researchers that have employed a more quantitative analysis, namely Lemme[8]. The use of a diffuser is common practice in many fluid applications, but its influence will never be greater than its impact on an exhaust system containing a catalytic converter. The flow as stated is not of a uniform velocity prior to reaching the front face of the catalyst, hence the angle of the diffuser should play an important role in its redistribution.

The quantity of air flow that passes through the catalyst must be the same quantity that is measured upstream using the hot wire meter, hence the continuity equation may be used in this analysis. The mass rate of flow is defined in EQ(5).

$$\dot{M} = \rho . \text{Area} . v \qquad \text{EQ(5)}$$

Therefore by assuming that density is constant, due to the relatively low flow rates and use of ambient air under steady state conditions, the volumetric flow rate may be calculated EQ(6).

$$\dot{Q} = \text{Area} . v \qquad \text{EQ(6)}$$

Then assuming that the flow is axisymmetric the total flow through the catalyst may be calculated and compared to the predicted and actual values from experimental results. In integrating EQ(7) it is assumed that the catalyst has a round cross section, Fig (7).

$$\int_0^Q d\dot{Q} = \int_0^A v \, dA \qquad \text{EQ(7)}$$

as $dA = 2.\pi.r.dr$

$$\dot{Q} = \int_0^R 2.\pi.r.v.dr \qquad \text{EQ(8)}$$

From the total flow through the catalyst EQ(8) it is possible to calculate an average velocity using the open area of the substrate. The open area being the frontal area of the substrate less the area of the cell walls.

$$v_{average} = \frac{\dot{Q}}{\text{Area}_{open}} \qquad \text{EQ(9)}$$

The average velocity as calculated in EQ(9), in itself, is ineffective in defining the ability of the diffuser to produce an even distribution of flow prior to the catalyst. Therefore to quantify the flow dispersion of the diffuser the average velocity must be related to the velocity profile which the catalyst substrate is subjected to. This can be done by using a regression analysis, so that the velocity distribution of the diffuser may be related mathematically to the radial distance from the central axis. A typical regression may be of the order 3 or 4, however this level of regression does not lend itself to a simple solution.

$$y = a_o + a_1 X \qquad \text{EQ(10)}$$

$$y = a_o + a_1 X + a_2 X^2 \qquad \text{EQ(11)}$$

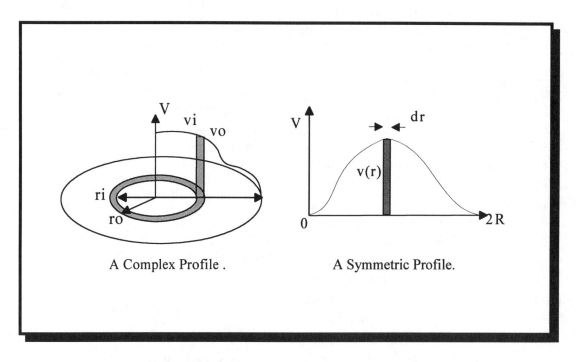

Fig 7. Mathematical Treatment of Flow Profiles

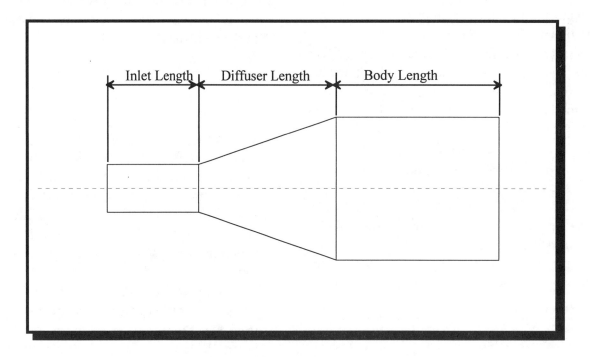

Fig 8. Geometry of Diffuser Canister

A regression of the order seen in EQ(10) was used for this analysis, although being of first order the result was not severely impaired. Therefore permitting the regression to take the form of EQ(12a) and using the substitutions indicated, a critical radius may be derived, EQ(12b). The critical radius is the point at which flow equates to the average velocity as calculated from EQ(1) and also indicates the boundary of where the catalyst is overflowed beyond its expected designed velocity.

$$v = a_o + a_1 R \qquad \text{EQ(12a)}$$

Substituting for ;

$$v = v_{average}$$
$$R = R_{critical}$$

$$R_{critical} = \frac{v_{average} - a_o}{a_1} \qquad \text{EQ(12b)}$$

The area enclosed by this radius establishes a definite point in the flow distribution at the substrate face at which the quantity of flow is at a transition point. The ratio of the excessively flowed area and the open area EQ(13) of the substrate denotes the use of area, a value close to one, indicating that the average flow velocity is distributed over the full face of the substrate.

$$Area_{effective} = \frac{Area_{critical}}{Area_{open}} \qquad \text{EQ(13)}$$

Therefore the above ratio denotes by what extent the frontal area is being flowed beyond its designed space velocity. The effective area ratio does not however provide sufficient information pertaining to what extent the critical area endures an increased mass flow. For example is the velocity of the gas through the critical area just slightly greater than the designed average velocity of flow, or is the critical area exposed to an above average flow concentration? Hence, the following series of equations, sum up the volumetric flow which would be incident on the $Area_{critical}$ and estimates if the quantity of flow is in proportion to the area contained within the

critical radius, EQ(14),(15),(16).

$$\dot{Q}_{critical} = \int_0^{R_{critical}} 2.\pi.v.r \, dr \qquad \text{EQ(14)}$$

$$\dot{Q}_{effective} = \frac{\dot{Q}_{critical}}{\dot{Q}} \qquad \text{EQ(15)}$$

From the above points it is possible to establish the flow distribution coefficient EQ(16).

$$FlowDist_{coeff} = \left[\frac{\dot{Q}_{effective}}{Area_{effective}} \right]^{-1} \qquad \text{EQ(16)}$$

This quantifies the capability of the diffuser to supply a uniform flow of air to the inlet of the catalyst substrate.

Ratio	Values	Comments
$Area_{effective}$	0.5< d >1	Good usage of frontal area 1. Slow warmup. 2. Good durability.
$Area_{effective}$	0< d >0.5	Concentrated area of flow 1. Fast warmup. 2. Poor durability.
$Q_{effective}$	0.5< d >1	Large flow concentration Flat faced velocity profile.
$Q_{effective}$	0< d >0.5	Low flow concentration Graduated velocity profile.
Flow Dist$_{coeff}$	0.5< d >1	Distribution of even flow across substrate face. Approaching optimum preformance. 1. Promotes durability
Flow Dist$_{coeff}$	0< d >0.5	Flow distribution uneven or highly concentrated. Substrate exhibits variety of flow rates across face. 1. Performance effected 2. Durability an issue.

Table 3. General Summary of Ratio Values

A value close to one indicates that the open flow area of the substrate has an incident flow profile which distributes a proportionate amount of flow to the critical area. A value less than one indicates that the flow area has a constrained flow which has an above average flow per unit area. Summarised in Table 3. are general comments related to each ratio and the implication of the calculated values obtained.

EXPERIMENTAL RESULTS

The flowing of the various geometries gives rise to the following results, Fig(9)-Fig(11). Observing the Reynolds number at the inlet to the geometry, typical values indicate that the flow is quite turbulent. This would suggest that the flow distribution would tend to be plug like in profile, Table 4.

As expected the shallow diffuser of Geom. 1 maintains the profile from the inlet pipe. As the flow increases, the profile is maintained with the air flow distribution in this case, being quite uniform across the catalyst. Inlet diffusers with large angles show a decrease in flow distribution across the catalyst.

Vol. Flow (m^3/s)	Inlet Pipe Reynolds	Catalyst Body Reynolds No.
8.5	7,816	3,908
16.1	14,804	7,402 .
27.1	24,913	12,456

Table 4. Reynolds numbers for Inlet Geometry

The severity of the flow distribution can be seen in Fig (11). The reason for such a deterioration in flow stems from the increase of diffuser angle to such a point that the flow is experiencing a sudden expansion into the catalyst housing. The presence of a tapering angle on the inlet of the diffuser permits the flow to stay attached to the wall thus promoting continuation of the flow profile created in the inlet pipe. In addition, increasing the inlet angle gives rise to the presence of a critical area on the catalyst which contains quite a high flow concentration. The nature of this flow pattern is due to the influence of the inlet diameter prior to the diffuser. As the flow rate increases then the ability of the flow to diffuse tends to decrease hence the efflux of gas

is still concentrated in a diameter shaped plume, Fig(12). Diffusers of this design will therefore create areas of high space velocities, thus promoting breakthrough under actual running conditions. The ability of the diffusers to generate an even flow distribution can be assessed by viewing the results of Fig(13). The inlet diffuser with the smallest angle, Geom.1 with an angle of 10° proves to give the best distribution of flow with Geom.3 proving to provide the worst distribution of flow.

PART II

Adding Empiricism to a Mathematical Model

THEORY

In reference to the paper presented by Carberry and Douglas[4] to this Society several assumptions were made as to the nature of the flow in the inlet section of the catalyst.

The above investigation into the influence of the inlet geometry on the flow through the catalyst substrate has shown that the assumption made, "that there was no flow maldistribution on the catalyst substrate", to be quite invalid. A rising from the information gleaned, this part of the model can now be ratified to reflect a closer correlation to what happens to real flows. Essentially the model determined its flow rate through the catalyst from EQ(1) coupled with EQ(4). This assumed that the flow was uniform over the inlet face of the catalyst.

At this juncture two distinct approaches can be made in ratifying the above initial assumption.

1. Typically the flow does not extend to the full circumference of the catalyst hence by decreasing the open diameter of the catalytic substrate fewer cell passages would be used for conversion within the algorithm, thereby lessening the efficiency of the catalyst.

2. From the quantity of flow determined to pass through the Area$_{critical}$ it is possible to recalculate the average velocity and use this value in the algorithm. However this would

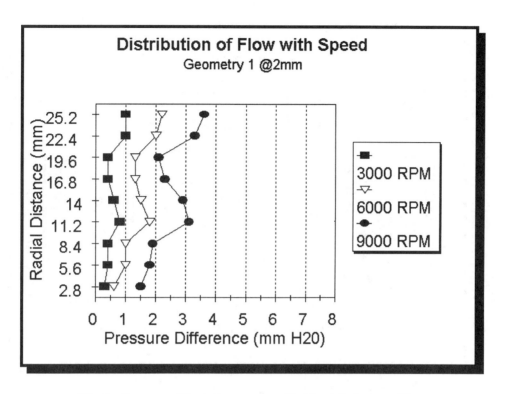

Fig 9. Pressure Variation across Catalyst Substrate Face
Due to 10 Degree Inlet Cone.

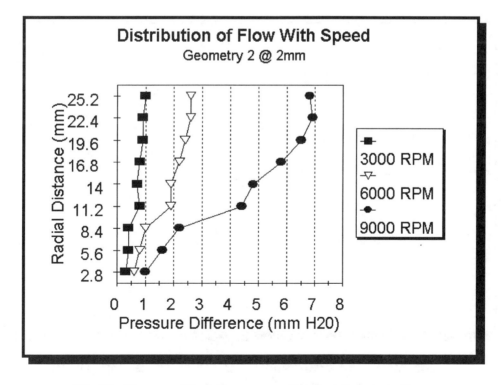

Fig 10. Pressure Variation across Catalyst Substrate Face
Due to 15 Degree Inlet Cone.

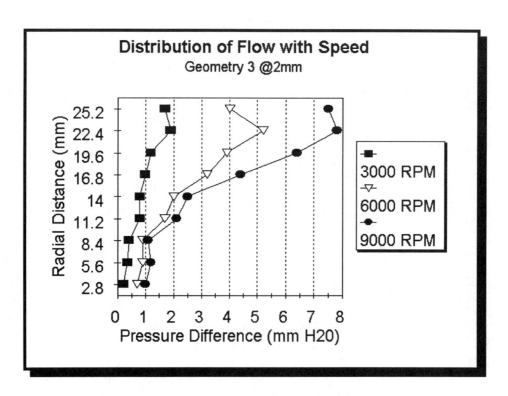

Fig 11. Pressure Variation across Catalyst Substrate face
Due to 45 Degree Inlet Cone.

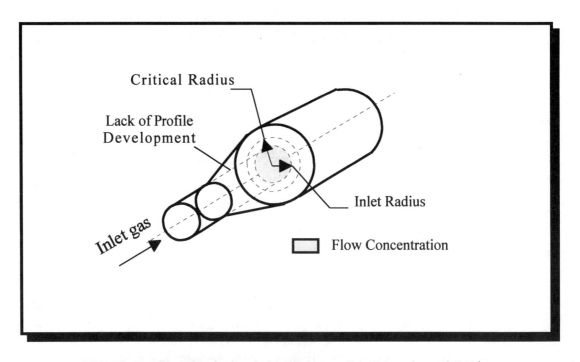

Fig 12. Profile Development leading to a Concentration of Gas

also require an assessment of the flow incident on the remaining catalyst area to be made. Hence the model is effectively carrying out two calculations, one on a high velocity area of catalytic substrate and then a calculation on a low velocity area.

Both of the methods described are meritorious. However in both cases the continuity of the flow is negated and to overcome this certain boundary conditions prevail for each application.

The first solution discussed above known as " Flow Distribution Model 1" (FDM1), requires the open diameter to be decreased to compensate for the lack of flow at the outer diameter of the catalyst and the increased flow in the centre. However in doing this the total flow through the catalyst is affected with the mass flow continuity not being conserved. An assumption may be made stating that the flow through the catalyst is increased by the same fraction that decreases the flow area., thus mass continuity is conserved.

For continuity

$$\dot{Q}_{before} = \dot{Q}_{after} \qquad EQ(17)$$

$$\dot{Q}_{before} = Area_{open} \cdot v_{average} \qquad EQ(18)$$

$$Area_{catalyst} = Area_{effective} \cdot Area_{open} \qquad EQ(19)$$

$$v_{catalyst} = v_{average} \cdot [Area_{effective}]^{-1} \qquad EQ(20)$$

$$\dot{Q}_{catalyst} = v_{catalyst} \cdot Area_{catalyst} \qquad EQ(21)$$

Thus

$$\dot{Q}_{before} = \dot{Q}_{catalyst} = \dot{Q}_{after} \qquad EQ(22)$$

The above treatment of the flow to be analysed in the catalyst model provides a much smaller area for the gases to flow through but also compensate for the increase in velocity that should also take place.

The second proposed method known as "Flow Distribution Model 2" (FMD2) to compensate for the flow distribution across the catalyst face is based on deducing two flow areas and two flow velocities. Combining EQ(13) and EQ(14) the average velocity may be determined for the $Area_{critical}$. A long with a similar determined value of velocity for the remaining area the mass continuity EQ(17) may then be applied to the system.

$$v_{critical} = \frac{\dot{Q}_{critical}}{Area_{critical}} \qquad EQ(23)$$

$$\dot{Q}_{remain} = \dot{Q}_{before} - \dot{Q}_{critical} \qquad EQ(24)$$

$$v_{remain} = \frac{\dot{Q}_{remain}}{Area_{open} - Area_{critical}} \qquad EQ(25)$$

$$\dot{Q}_{catalyst} = \frac{Area_{critical} \cdot v_{critical} + Area_{remain} \cdot mf \cdot v_{remain}}{} \qquad EQ(26)$$

A term *mf* is included in the remaining equation EQ(26), this term ensures that continuity is preserved before and after the catalyst. This method ensures that two distinct flow rates are represented in the catalyst model.

COMPUTATIONAL RESULTS

The two models described above were used to modify the flow characteristics of the inlet flow to the catalyst substrate. The above models of the flow were included in the catalytic converter simulation model. The same inlet conditions were applied to the model which simulated the flow from the catalyst rig Carberry et al [4]. The results from the applied theory can be seen in Fig(14) and Fig(15). Essentially the correlation is much closer to the measured results using the modifier of the flow than assuming :

1. The full catalyst inlet area in used.
2. The inlet velocity is uniform across the catalyst substrate inlet.

The two modifiers of the inlet flow increase the

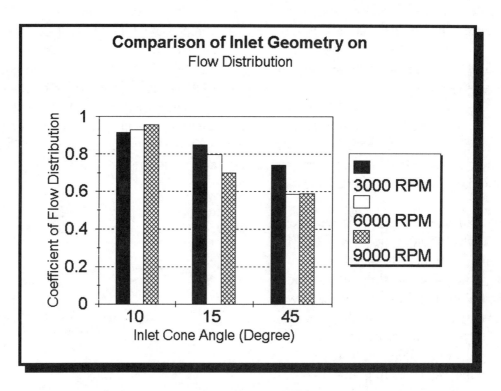

Fig 13. Coefficient of Flow Distribution for Inlet Cones

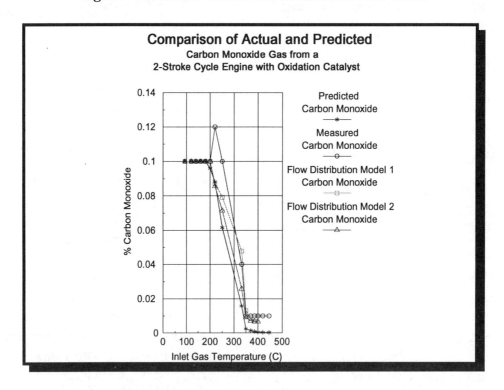

Fig 14. A Comparison of Different Inlet Boundary Models and their affect on Carbon Monoxide Prediction.

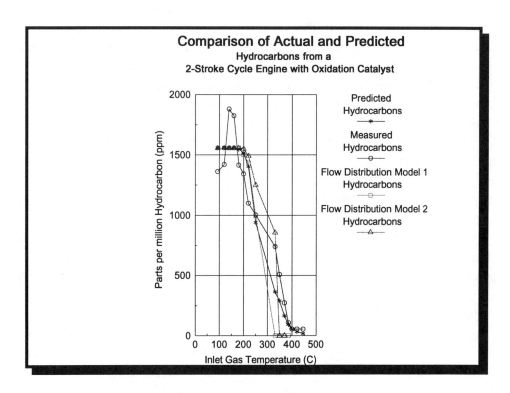

Fig 15. A Comparison of Different Inlet Boundary Models
and their affect on Hydrocarbon Prediction.

accuracy of the prediction but at the expense of an increase in computational time. To analyse the results it is necessary to understand that the conversion efficiency of the catalyst is firstly; a function of the flow rate through the catalyst and secondly a function of the gas to be oxidised. With this information at hand, the prediction of the presence of carbon monoxide CO can be analysed. Firstly, flow distribution model 1.(FDM1) predicts the conversion of the gas at the temperature observed quite well for the mid range of inlet gas temperatures, 290°C to 330°C. The algorithm used promotes the use of one inlet gas velocity on a decreased substrate area, this model represents a close resemblance to what happens in reality. However, the fact that the model can predict the presence of CO quite well actually belies the reason why the model does not predict the hydrocarbon concentration as accurately. As a result of the decrease of oxygen due to the oxidation of the CO the hydrocarbons are left only partially oxidised.

Flow distribution model 2.(FDM2) suffers generally from over prediction. Although the values predicted for CO and HC are again better than those of the standard predictive model. The use of two flow areas, greatly increases the amount of reactive surface available for reaction therefore promoting an increase in the number of reactions occurring per unit area. This model is ultimately suited to higher flow rates, where the two flow areas due to the division of flow permits a larger than average velocity to be quantified for the critical flow area. In addition to a lower than average velocity to be calculated for the peripheral substrate area. As stated earlier one of the factors affecting the conversion taking place is the rate of flow, a flow velocity lower than design velocity promoting a good rate of reaction and vice versa.

SUMMARY AND CONCLUSIONS

A problem exists with the inability of catalytic converters designed for a specific application to fully function. This was directly a result of the incorrect design of inlet cones and routing of pipe work.

The experimental work performed showed that flow distribution was not only a function of inlet cone design but also the rate of flow.

A method has been devised that assesses the ability of an inlet cone to distribute flow adequately for the catalytic converter to function correctly.

The information obtained from the experiments were also used to develop a flow distribution model for a simulation program of a catalytic converter. The two models developed took account of the flow distribution using two distinct methods. A comparison of predicted and measured data show that the FDM 1 algorithm accounted well for the flow distribution being suited generally to comparatively low flow rates.

ACKNOWLEDGEMENTS

The authors would like to express their gratitude to the Queen's University of Belfast for providing the facilities and expertise. The following Companies are also recorded a note of thanks for supplying the substrates for testing, Johnson & Matthey and Emitec. Finally we would like to express our gratitude to the Department of Education for Northern Ireland for supplying financial support.

ABBREVIATIONS

CO   Carbon Monoxide.
HC   Hydrocarbons.
FDM   Flow Distribution Model.

NOMENCLATURE

d	Arbitrary ratio value	
SP.Vel	Space velocity	hr^{-1}
B	Engine capacity per hour	m^3/hr^{-1}
CV	Catalyst Volume	m^3
a_o	Regression Coeff.	-
a_1	Regression Coeff.	-
$Res_{(TIME)}$	Residence time	hr
P_{TOTAL}	Total Pressure	N/m^3
P_{STATIC}	Static Pressure	N/m^2

Symbol	Description	Units
ρ	Density	kg/m^3
mf	Coefficient	-
V	Velocity axis	m/s
V_{remain}	Model 2 velocity Peripheral area	m/s
$V_{average}$	Model 1,2 velocity Open area	m/s
v_i	Velocity inner	m/s
v_o	Velocity outer	m/s
r_i	Radius inner	m
r_o	Radius outer	m
r	Radius	m
R	Radius axis	m
$\dot{M}$	Mass flow rate	kg/s
$\dot{Q}$	Volumetric flow rate	m^3/s
$\dot{Q}_{before}$	Volumetric flow before catalyst	m^3/s
$\dot{Q}_{after}$	Volumetric flow after catalyst	m^3/s
$\dot{Q}_{catalyst}$	Volumetric flow rate (continuity)	m^3/s
$\dot{Q}_{remain}$	Volumetric flow Model 2 peripheral area	m^3/s
$\dot{Q}_{critical}$	Volumetric flow critical area	m^3/s
$\dot{Q}_{effective}$	Ratio	-
Area	Frontal area	m^2
$Area_{open}$	Open area catalyst	m^2
$Area_{critical}$	Critical area catalyst	m^2
$Area_{effective}$	Ratio	-
$Area_{remain}$	Remaining area Model 2	m^2
$FlowDist_{coeff}$	Coefficient	-

REFERENCES

[1] M. L. Church, B. J. Cooper, P. J. Wilson, " Catalyst Formulations 1960 to Present", International Congress and Exposition Detroit, Michigan Feb27 - Mar3, 1989, SAE Paper No. 890815

[2] S. T. Gulati, J .C. Summers, D. G. Linden, J. J. White, "Improvements of converter durability and activity via Catalyst Formulation", International Congress and Exposition Detroit, Michigan Feb27 - Mar3, 1989, SAE Paper No. 890796

[3] S. O. Oh , C. C. Eickel ," Effects of Cerium addition on CO Oxidation Kinetics over Alumina- Supported Rhodium Catalysts", Journal of Catalysis Volume 112, No. 2, August 1988.

[4] B. P. Carberry, R. Douglas, "A Simple But Effective Catalyst Model for Two-Stroke Engines". International Off-Highway and Powerplant, Congress and Exposition, Milwaukee, Wisconsin Sept 27-3, 1992, SAE Paper No. 921693.

[5] D. W. Wendland, W. R. Matthes, "Visualization of Automotive Catalytic Converter Internal Flows", International Fuels and Lubricants Meeting and Exposition Philadelphia, Pennsylvania Oct 6-9, 1986, SAE Paper No. 861554

[6] A. J. Baxendale," The Role of Computational Fluid Dynamics in Exhaust System Design and Development",1993 Vehicle Thermal Management Systems Conference Proceedings P-263, SAE Paper No.931072.

[7] M. C. Lai, J. Y. Kim, C. Y. Cheng, P. Li, G. Chui, J. D. Pakko, " Three Dimensional Simulations of Automotive Catalytic Converter Internal Flow. SAE Paper No. 910200.

[8] C. D. Lemme, W. R. Givens, " Flow Through Catalytic Converters - An Analytical and Experimental Treatment", Automotive Engineering Congress Detroit, Michigan Feb25 - Mar1, 1974, SAE Paper No. 740243.

The Initial Development of a
Two-Stroke Cycle Biogas Engine

K. Doherty, Gordon P. Blair, R. Douglas, and R. Kee
The Queen's University of Belfast

J. Purdy
Dept. of the Environment for N. I.

ABSTRACT

Anaerobic digestion is a popular method of treating sewage sludge. Biogas or sewage gas is a by-product of this process. Significant volumes of biogas are produced at many sewage treatment works and also at some landfill sites from the natural breakdown of municipal waste. This biogas can be used as a fuel for an engine and generating set, producing electrical power and heat.

A multi-cylinder two-stroke cycle system, capable of being retrofitted to current production four-stroke cycle engines, is proposed, primarily for the combustion of biogas in combined heat and power applications. The engine incorporates features to give good tolerance to the corrosive agents associated with biogas. This paper describes the design and initial development of a purpose built single cylinder research engine to investigate this concept. A low pressure direct injection system which has been developed for use with the engine is also outlined. Finally, some results from the initial running of this engine on actual biogas are presented.

1.0 INTRODUCTION

Anaerobic digestion is one of several treatment processes used to reduce the biological activity of sewage sludge. It is a bacterial process in which organic material is decomposed in the absence of oxygen. Methanogenesis is the final stage of this complex biological process. Here products of earlier stages are utilized by methanogenic bacteria to produce methane and carbon dioxide, the main constituents of biogas. This gas consists typically of 60% - 70% by volume methane, with the balance comprising of carbon dioxide and a small proportion of hydrogen sulphide. Significant volumes of biogas are produced in many sewage treatment works and also at some landfill sites from the natural breakdown of municipal waste.

Biogas is a fuel that can easily be used in a boiler or a kiln. Alternatively, the gas can be used to fuel an internal combustion engine, producing power and heat in a co-generation or combined heat and power (CHP) installation. The power output can be used to generate electricity, while the heat can be used for a number of applications, including maintaining the digester at its optimum operating temperature in a sewage treatment works. By utilizing such as system, 75% - 80% of the energy available in the biogas can be realized [1]. The commercial viability for such combined heat and power systems is dependent on the initial installation costs, running costs and the expected energy output value. A significant element in the running costs, especially with small-scale CHP systems, is associated with engine maintenance and engine component replacement. The reliability and durability of the engine is therefore of primary importance.

Biogas is an adequate fuel for conventional internal combustion engines. However the presence of hydrogen sulphide, in varying concentrations from 100 - 10000 ppm, depending on the raw nature of the sewage sludge, introduces significant reliability and durability problems. The hydrogen sulphide gas, even in relatively small concentrations, is particularly corrosive, attacking copper based bearing materials that are predominantly used in internal combustion engine design. Picken et al [2], considered the main cause of engine failure, for conventional engines operating on biogas, to be the blow-by of unburnt charge past the piston rings and valve guides. The hydrogen sulphide gas, present in the blow-by charge, was then corroding vital engine components in its gaseous state as well as contaminating the lubricating oil. Although the resulting acidification of the lubricating oil may be controlled with careful lubricant choice and regular oil changes, the prevention of corrosion by the hydrogen sulphide vapour is a more difficult problem to overcome. Indeed, in a study carried out by Fulton [3], it was concluded that the economics of CHP in sewage treatment works should be seriously questioned at locations where the hydrogen sulphide levels in the biogas would exceed 2000 ppm. Gilbert et al [4] reinforced this view, concluding that where hydrogen sulphide levels exceeded 1000 ppm some form of gas treatment was necessary before it could be used as an engine fuel. However, despite this problem of the blow-by of hydrogen sulphide gas, economic features have essentially dictated that most current biogas CHP installations are based on adapted models of conventional four-stroke liquid fuel burning engines, running on untreated biogas.

This paper describes the design and initial development of an alternative power unit specifically suited to the combustion of biogas in a CHP application. The engine design includes features that provide good tolerance to the corrosive agents associated with biogas, and thus will enable the engine to operate efficiently on biogas regardless of the level of these corrosive agents.

2.0 ENGINE DESIGN

2.1 GENERAL - The proposed system comprises a two-stroke cycle cylinder and crosshead assembly. This system can be retrofitted to current multi-cylinder production four-stroke cycle engines. A single cylinder research engine of a swept volume of 432 cm³ was designed to investigate this concept. The engine is a crosshead type, piston ported two-stroke design and is shown in Figure 1. This design allows the use of a conventional production four-stroke crankcase assembly, while also providing a barrier to the corrosive biogas agents. The engine is designed to operate continuously at 1500 rpm, as is normal in a CHP application, and as such utilizes a fully tuned exhaust system to give peak torque at this speed. The engine specification is given in Table I.

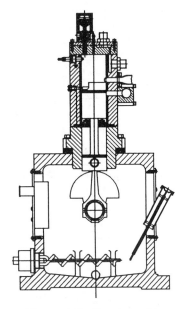

Figure 1 Engine Assembly

The two-stroke design has some inherent advantages over its four-stroke counterparts for this particular application. The traditional two-stroke engine is mechanically simple with no need for expensive overhead valve gear or a recirculating lubricating system. It therefore avoids many of the problem areas associated with four-stroke engines operating continuously on biogas. The two-stroke engine also has the potential for a higher specific power output when running on biogas, with a power stroke for each crankshaft revolution. The equivalent four-stroke engine only generates a power stroke for every two crankshaft revolutions. This increased firing frequency would also produce a smoother power source better suited for electricity generation. A further advantage of the two-stroke engine is its inherently low NO_x emissions, important with increasing legislation directed at stationary power generating engines.

2.2 ENGINE OPERATION - The crosshead

Table I Engine Specification

Induction System	Piston Port
Bore	80 mm
Stroke	86 mm
Displacement	432 cm³
Trapped Compression Ratio	10.2:1
Exhaust Port Opens	123° atdc
Transfer Port Opens	138° atdc
Inlet Port Opens	45° btdc
"Crankcase" compression Ratio	1.62:1
Inlet Diameter	27.5 mm
Exhaust System	Fully Tuned

two-stroke design consists of two pistons running in separate bores. The upper piston or 'power piston' and barrel operate as a conventional two-stroke internal combustion engine, whereas the lower piston or 'crosshead' is essentially a guide through which power, generated in the power cylinder is transmitted to the crankshaft. The power cylinder is sealed from the crosshead and standard crankcase assembly by a crosshead seal. The biogas is supplied directly to the upper or power cylinder, well away from the crankcase. The volume between the underside of the power piston and the crosshead seal is used to pump pure air into the power cylinder in the normal two-stroke fashion. Fresh air is induced into this volume with each up-stroke of the piston. This fresh air then scavenges the under piston pumping volume of any biogas or combustion products, which may have leaked past the power piston, returning them to the power cylinder. Moreover, the pressure difference between the under piston pumping volume and the crankcase is small, therefore, blow-by into the crankcase is greatly reduced and will consist mainly of pure scavenge air. By this method, the concentration of biogas and biogas combustion products reaching the crankcase is significantly reduced.

Lubrication of the crankshaft assembly continues with the original automotive lubrication system, which is never exposed to any of the biogas contaminants. Lubrication to the power cylinder is of the total loss method, normally adopted with two-stroke engines, where a small quantity of oil is metered into the inlet manifold and carried by the scavenging air into the cylinder. This will also limit the build-up of any corrosive agents within the power cylinder.

2.3 COMBUSTION SYSTEM - The power cylinder adopts the two-stroke cross scavenging principle and employs the unique QUB deflector piston. This piston design, developed at QUB and described with test results by Blair et al [5], is of particular interest and can be seen in Figure 2. The flat crown and compact combustion chamber, in conjunction with the flat cylinder head, gives rise to high squish velocities and hence high turbulence within the combustion chamber. High turbulence and effective scavenging of the spark plug gap were indicated by Kingston Jones et al [6] to be important parameters for the efficient combustion of biogas. Indeed Kee et al [7] demonstrated that a combustion system of this type produces a fast combustion process, which is ideal for the burning of biogas, with its naturally low laminar burning velocity.

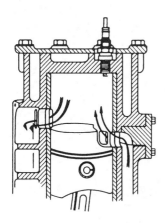

Figure 2 QUB Cross Scavenging

The QUB cross scavenging principle also allows the piston to run cooler than conventional raised deflector pistons. Furthermore, as the piston rings are not exposed to the hot exhaust gases during the exhaust blowdown phase, oil consumption levels are reduced and any potential for ring sticking is limited.

The use of a cross scavenged type two-stroke design has also some considerable packaging and manufacturing advantages over a possible loop scavenged design. The compact scavenging port layout of the cross scavenged design, shown in Figure 3, allows the closest possible cylinder to cylinder spacing in a multi-cylinder configuration. Additionally, the machined ports, normally employed with a QUB cross scavenged engine, will ensure cylinder to cylinder consistency in a fashion not possible with the somewhat more expensive casting processes required for a loop scavenged engine in mass production.

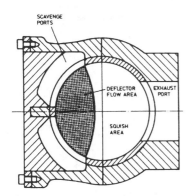

Figure 3 Port Layout

2.4 FUEL DELIVERY SYSTEM - A major disadvantage of the two-stroke engine has historically been the loss of fuel to the exhaust during scavenging of the cylinder. This has resulted in the two-stroke engine being characterized as an engine with poor fuel economy and high hydrocarbon emissions. However advances in scavenging and stratified charging techniques have gone someway to alleviate this notoriety.

Stratified charging involves arranging for the supply of fuel, or fuel and air, to be held in one area of the cylinder, while the air scavenging process, expelling the exhaust gases, takes place in a different area of the cylinder. Effectively, a barrier of pure air, or at least a very lean mixture, is placed between a fuel rich region and the exhaust port. This technique can increase significantly the efficiency of fuel retention in a two-stroke internal combustion engine.

As such, it was envisaged that some form of stratified charging would be needed with the two-stroke crosshead biogas engine, to improve fuel retention in the cylinder and in conjunction with the QUB cross scavenging system, enhance combustion within the cylinder.

One stratified charging technique that can be employed with a two-stroke engine is that of direct fuel injection. In a direct fuel injection system, only pure air is used to scavenge the cylinder of exhaust gases. Fuel is then injected directly into the cylinder just after the exhaust port has closed or shortly before it has done so. In recent years, developments in the automotive world have focused attention on direct injected two-stroke engine designs, as a means to meeting future automotive performance requirements. Advances in this area are well documented and much experience has been accrued at QUB with gasoline direct injected two-stroke engines [8, 9]. However, development of gas type fuel injectors is not as well advanced, although some experience has been gained at QUB [10].

In a conventional direct injection gasoline system, an injector must deliver the required quantity of fuel at a timing sufficient to allow the fuel to distribute, vaporize and mix with the air before combustion is initiated. The time available for such a process, with a two-stroke engine, is very limited. With a gaseous fuel, although the vaporization period is eliminated, larger quantities of a less dense fuel must now be delivered to the cylinder. Taking these requirements into account, Green et al [11] demonstrated that a natural gas direct injection system could be developed successfully for use on a two-stroke engine.

It was thus decided that direct injection of biogas in this particular application was one method of fuelling that would have to be investigated. This necessitated the development of a gas injector suitable for biogas delivery that would be unaffected by its corrosive nature. It was thought that the injector would be of a low pressure, solenoid actuated design, similar to that outlined by Green et al [11] in their report. As a result, an injector for use on the two-stroke crosshead biogas engine was designed within QUB [12].

The injector arrangement is shown in Figure 4 and incorporates a solenoid actuated poppet valve to control the injection of a low pressure biogas supply. It is essentially a gas metering valve, where a constant pressure biogas supply is metered to the engine. The injector operation is illustrated in Figure 5. The biogas is supplied to the annular volume around the poppet valve stem. The poppet valve is held close on its seat against this gas pressure by the action of a preloaded spring. When the solenoid is activated, the poppet valve is opened and the biogas enters the combustion chamber under its own pressure. The poppet valve lift is limited when the moving armature of the solenoid contacts the body of the solenoid. This maximum lift value can be varied, as can the spring preload which allows the use of different supply gas pressures. The solenoid is isolated from the gas supply by a seal to prevent possible contact with any biogas corrosion agents. The injector specification is shown in Table II.

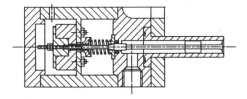

Figure 4 QUB Biogas Injector

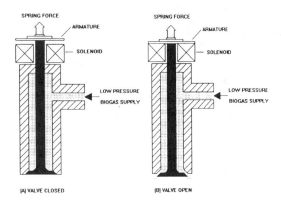

Figure 5 Schematic of Injector Operation

Table II Injector Specification

Solenoid	Lucas 1EC
Valve Diameter	10 mm
Valve Lift	0.8 mm
Spring preload	21.7 N

3.0 EXPERIMENTAL TESTING PROGRAMME

A firing engine test programme was carried out on the QUB biogas engine with direct fuel injection. For these tests the biogas injector was located in the cylinder head and positioned centrally over the combustion chamber. The research engine can be seen mounted on the test bed in Figure 6.

3.1 APPARATUS - The test engine was directly

Figure 6 QUB Biogas Engine

coupled from the crankshaft to a Borghi and Saveri eddy-current type dynamometer. The biogas was supplied to the engine from regulated high pressure gas cylinders, filled to a pressure of 160 bar at a neighbouring sewage works. Each supply of biogas was analyzed to determine its quality. The biogas consisted typically of 60% (by volume) methane, 36% carbon dioxide and 4% water vapour, with a hydrogen sulphide content typically in the range 100 - 300 ppm. A full specification for the biogas used during engine testing is given in Table III. Gas flow rate was measured using a Fischer & Porter variable-area rotameter and the air consumption rate was determined from the pressure drop across a BS1042 [13] orifice plate which admitted air to a large baffled surge tank.

Table III Biogas Specification

Methane	58.6% by volume
Carbon Dioxide	37.7%
Nitrogen	0.2%
Water Vapour	3.5%
Hydrogen Sulphide	70 ppm
Net Calorific Value	17.6 MJ/kg
Stoichiometric AFR	6.0 (by mass)

An Oliver Multigas exhaust analyzer was used for emissions analysis. This analyzer employs heated FID detection of HC, NDIR for CO and CO_2, Chemiluminescence for NO_x, and a paramagnetic system for O_2. The air and fuel trapping efficiencies, trapped air/fuel ratios and the brake specific emissions were calculated as described by Douglas [14].

3.2 TEST PROCEDURES - As the engine is intended primarily for use in a CHP installation, all of the tests were carried out at wide open throttle and full load, at a speed of 1500 rpm.

3.2.1 Baseline Testing - A baseline test was conducted with a low pressure fuel supply to the inlet manifold, to determine the performance and emission characteristics of the engine when scavenged with a homogeneous mixture. The air/fuel ratio was varied between the rich and lean misfire limits. At each test point, the ignition timing was adjusted to the minimum advance for best torque (MBT).

3.2.2 Direct fuel injection - Tests were conducted with the biogas injector at nominal start of gas injection timings of 180° and 200° atdc. The supply gas pressure to the injector was maintained at a nominal value of 345 kPa for each of these tests. At each injection timing an air/fuel ratio sweep was carried out. The ignition was set at the MBT value for every test point. The direct injection tests are presented and compared with the baseline results in Figure 7.

4.0 DISCUSSION OF RESULTS

The performance results show that for comparable trapped air/fuel ratios, the direct injection of biogas produced significant improvements in brake thermal efficiency and hydrocarbon emissions, over those obtained with homogeneous mixture scavenging. The maximum brake thermal efficiency was increased from 20.2% at a bmep of 373 kPa, for homogeneous mixture scavenging, to 25.6% at a bmep of 403 kPa, for direct injection at 200° atdc. The maximum bmep points were 389 kPa for the homogeneous mixture and 468 kPa for direct injection at 200° atdc, resulting in brake thermal efficiencies of 18.6% and 24% respectively. Corresponding to these maximum bmep points, methane consumption (as opposed to biogas consumption) was 372 g/kWh and 298 g/kWh respectively.

The stoichiometric air/fuel ratio for the biogas used in the testing programme was 6.0. The maximum brake thermal efficiencies stated above were obtained at trapped air/fuel ratios of 6.3 for homogeneous scavenging and 7.1 for direct injection at 200° atdc. The maximum bmep points however were obtained with similar trapped air/fuel ratios of 5.7 and 5.5 respectively. From the results it is also clear that there was a distinct narrowing of the flammable air/fuel ratio limits with direct fuel injection. The bmep of 468 kPa recorded with the test engine is equivalent to 936 kPa for a four-stroke engine and is exceptional considering the low quality fuel being used. This can be compared with a typical bmep of about 650 kPa for normally aspirated four-stroke spark ignited gas engines operating on biogas. However care should be taken when making such comparisons as the performance of engines of this type is very dependant on both fuel quality and engine operating conditions.

The bsHC emissions were reduced by a minimum of 77% across the air/fuel ratio range with the adoption of direct biogas injection at 200° atdc. It should be noted at this point that the hydrocarbon emissions quoted, represent the total hydrocarbons emitted from the engine, and therefore include unburnt methane from the fuel. However, as methane does not react in the photochemical smog reaction, it is not generally considered a pollutant. Nevertheless, methane is a greenhouse gas and from an environmental point of view, its emission must be minimized.

The $bsNO_x$ characteristics, which are normally dependant on peak cylinder temperature and pressure, showed little change with the different methods of fuelling. For the maximum bmep points stated, $bsNO_x$ levels (expressed in NO_2 equivalent) were 5.4 g/kWh for homogeneous scavenging and 3.6 g/kWh for direct injection at 200° atdc. These values, expressed in terms as used by the German TA-Luft Stationary Engines Emissions Regulations, translate to 1.0 g/m^3 for homogeneous scavenging and 0.71 g/m^3 for direct injection at 200° atdc. The TA-Luft Regulations state a limit of 0.8 g/m^3 for a stationary two-stroke power generating engine. However it should be noted that at present there exists in the United Kingdom no current or proposed emissions legislation for stationary power generating engines burning gas derived from the biological degradation of waste [15]. Clearly engines operating on biogas fall into this particular group.

The fuel trapping efficiency was also improved with direct injection, with typically only 6% of the fuel being lost directly to the exhaust, as opposed to a minimum loss value of 31% obtained with homogeneous scavenging. The air trapping efficiencies however remained relatively unchanged with the employment of direct injection.

The delivery ratio, which is a measure of the engine's ability to intake air, was also increased under direct injection conditions. This is due to the fact that with homogeneous mixture scavenging, the biogas that is being supplied to the inlet manifold actually displaces some of the intake air. This has been observed with other homogeneously charged engines operating on gaseous fuels [16], and would account for some of the power increase experienced when the fuel is injected directly into the cylinder.

The MBT timings appeared to be unaffected with the adoption of the alternative fuelling methods. The MBT timings were typically in the range of 10° - 15° btdc, and would suggest that a rapid combustion process was being achieved.

Finally, an analysis of the crankshaft lubricating oil carried out by the Mobil Oil Company revealed minimal contamination with biogas corrosive agents after 50 hours of continuous engine operation. This would suggest that the biogas contaminates were indeed being prevented from reaching the crankshaft assembly and its lubricating system. However, it should be remembered that as normal oil degradation is usually considered over periods of 250 hours or more, these results after a period of only 50 hours of engine operation are not sufficient to allow any firm conclusions.

Further comparison of the engine's performance with those of four-stroke gas engines operating on biogas is limited due to the lack of relevant information published

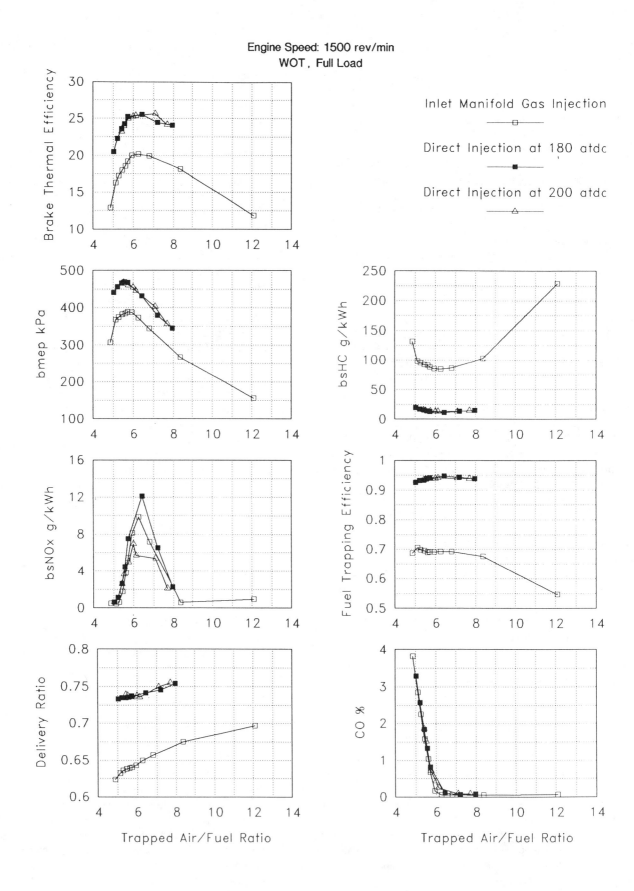

Figure 7 Comparison of Engine Test Results

in the literature. It is anticipated however that further development of this engine concept will involve testing a standard four-stroke spark ignited gas engine on biogas, conversion to the two-stroke principle and a full comparison made of the performance, emissions characteristics and oil degradation.

5.0 CONCLUSIONS

(1) Tests, conducted on the 432 cm^3 single cylinder crosshead two-stroke engine, have demonstrated that this engine concept does have potential as an alternative power unit for the combustion of biogas in a CHP application.

(2) Power outputs of 11.8 kW/litre were achieved at an engine speed of 1500 rpm, using biogas with a net calorific value of only 17.6 MJ/kg.

(3) The engine tests also showed that large improvements in engine power output, brake thermal efficiency and hydrocarbon emissions could be attained with the use of direct biogas injection in place of homogeneous charging.

(4) Maximum brake power output was increased by 21% with the adoption of direct biogas injection. This was primarily due to the increased delivery ratio obtained, as the biogas was no longer displacing air in the intake manifold, as is the case with homogeneous charging.

(5) At maximum bmep conditions brake thermal efficiency was increased from 18.6% to 24%, with a corresponding 20% reduction in brake specific gas consumption. This resulted in a methane consumption figure of 298 g/kWh, attained at a bmep of 468 kPa.

(6) Hydrocarbon emissions, were reduced by a minimum of 77% across the air/fuel ratio range with direct biogas injection. This was essentially due to a 80% reduction in fuel lost to the exhaust.

(7) It also appeared that a rapid combustion process was being achieved with the QUB cross scavenging system, with typical MBT timings in the range 10° - 15° btdc for either fuelling method.

(8) Initial results from the continual oil monitoring program for the crankshaft lubricating system suggest that the standard automotive type crankshaft assembly was not being exposed to any significant levels of biogas corrosive agents.

ACKNOWLEDGEMENTS

The authors wish to thank The Queen's University of Belfast for the provision of workshop and laboratory facilities. The technical staff are also acknowledged for their skillful work and valuable assistance during experimental testing. Thanks are also due to The Department of the Environment for Northern Ireland for their financial support of this research project.

REFERENCES

1   N. Ladommatos and C.R. Stone, "Conversion of a small Diesel Engine for Gaseous Fuel Operation at High Compression Ratio", SAE Paper No. 910849, February 1991.

2   D.J. Picken, M.F. Fox and G.P. Bush, "An Investigation into Piston Ring Blow-By and a Discussion of its Impact on Biogas Engines", IMechE C433/020, 1991.

3   A.C. Fulton, "The Effect of Hydrogen Sulphide on CHP Engines at Countess Wear Sewage-Treatment Works, Exeter", J.IWEM, April 1991.

4   R.J. Gilbert and D.B. Wootton, "The Lubrication of Gas Engines", Seminar on Gas Engines and Co-Generation, IMechE, May 1990.

5   G.P. Blair, R.A.R. Houston, R.K. McMullan, N. Steel and S.J. Williamson, "A New Piston Design for a Cross Scavenged Two-Stroke Cycle Engine with Improved Scavenging and Combustion Characteristics", SAE Paper No. 841096, 1984.

6   M.G. Kingston Jones and D.M. Heaton, "Nebula Combustion System for Lean Burn Spark Ignited Gas Engines", SAE Paper No. 890211, 1989.

7   R.J. Kee, G.P. Blair and R. Douglas, "Comparison of Performance Characteristics of Loop and Cross Scavenged Two-Stroke Engines", SAE Paper No. 901666, 1986.

8   G.P. Blair, R.J. Kee, C. Carson and R. Douglas, "The Reduction of Emissions and Fuel Consumption by Direct Air-Assisted Fuel Injection into a Two-Stroke Engine", 4th Graz Two-Wheeler Symposium, TU GRAZ, April 1991.

9   C. Carson, R.J. Kee, R. Kenny and G.P. Blair, "The Reduction of Exhaust Emissions from Two-Stroke Engines", Seminar on Worldwide Engine Emission Standards and How To Meet Them, IMechE, May 1993.

10   M. Donohoe, "The Design and Development of a Natural Gas Fuelled Two-Stroke Engine", Final Report No. 2004, Department of Mechanical and Manufacturing Engineering, The Queen's University of Belfast, February 1990.

11   C.J. Green, J.S. Wallace, "Electrically Actuated Injectors for Gaseous Fuels", SAE Paper No. 892143, 1989.

12   P. Davison and R. Rutledge, "Development of a Biogas Fuel Injector for a Two-Stroke IC Engine", Final Year Reports No. 2010 and 2011, Department of Mechanical and Manufacturing Engineering, The Queen's University of Belfast, February 1990.

13   BS1042, "Fluid Flow in Closed Conduits", British Standards Institution, 1981.

14   R. Douglas, "AFR and Emission Calculations for Two-Stroke Cycle Engines", SAE Paper No.

901599, September 1990.

15   T.O.R. Shaw, "The Law's Demands", Seminar on
        Stationary Combustion Engines and The
        Environmetal Protection Act, IMechE, December
        1992.

16   R.L. Evans, F. Goharian and P.G. Hill, "The
        Performance of a Spark-Ignition Engine Fuelled
        with Natural Gas and Gasoline", SAE Paper No.
        840234, February 1984.

932399

Non-Isentropic Analysis of Varying Area Flow in Engine Ducting

G. P. Blair and S. J. Magee
The Queen's University of Belfast

ABSTRACT

In two previous papers to this Society (1,2)* an 'alternative' method was presented for the prediction of the unsteady gas flow behaviour through a reciprocating internal combustion engine. The computational procedures led further to the prediction of the overall performance characteristics of the power unit, be it operating on a two- or a four-stroke cycle. Correlation with measurements was given to illustrate its effectiveness and accuracy.

In the ducts of such engines there are inevitably sectional changes of area which are either gradual or sudden. A tapered pipe is typical of a gradual area change whereas a throttle or a turbocharger nozzle represents a sudden area change. In those previous papers it was indicated that a fuller explanation, of the theoretical procedures required to predict accurately the unsteady gas flow in such duct sections would be given in a later paper to this Society; this is that necessary publication.

The theory of gradual and sudden area changes is presented, together with computational illustrations of its application to real geometrical cases. The theory includes non-isentropic effects at such area changes and inherently solves the mass continuity, energy and momentum equations at each section.

INTRODUCTION

There has been some history, extending over thirty years of study, at QUB on research into unsteady gas flow through the reciprocating internal combustion engine(3). Much of it was based on the pioneering computational work of Benson(4).

By the early 90's it was becoming clear that the computational methods were becoming ever more complex while accuracy was little better than it had been almost at the outset. New methods had appeared, using finite difference or volume schemes(5,6,7). Each of these new methodologies had their own particular advantages or disadvantages. However, all methodologies which solve computationally the unsteady gas flow regime, in pipes with gradual or sudden changes of area, by relying on terms which are not related specifically to denoting the flow as an expansion or a contraction process are doomed inevitably to inaccuracy. Where there are gradual changes of area in a duct, particularly for an exhaust process, the inability of any calculation method to recognise that high particle velocity flow separation is occurring, is also doomed to inaccuracy.

It is in recognition of these effects that this paper is written and computational proof of these assertions is presented. The fundamental

* Numbers in parentheses designate References at the end of the paper.

computational method has already been presented[1]. Within previous papers [2,8], where calculations of high performance two- and four-stroke engines were compared very favourably with experimental measurements, a newer technique of computing the varying area flow was employed. This paper provides the theoretical explanation of that improved method for determining the wave transmission in ducts which have a varying or a sudden area change with length.

1. THEORY

1.1 <u>The background to the proposed theory</u>. The tracking of the unsteady gas motion, for example in the method of characteristics, is carried out by a mesh calculation for characteristics called Riemann variables[4]. In the alternative method already proposed, the tracing of the pressure wave motion is also conducted by a meshing of the ducts in distance terms and following the motion of the leftward and rightward pressure waves in short time steps. All of this has already been presented in some considerable detail[1]. It is not proposed to change that procedure, except in one regard, and that is for the calculation of the reflection of these pressure waves as they pass through ducts of varying area or encounter sudden changes of section. To do that efficiently requires a subtly different manner of meshing the duct system, and this too will be described herein. In most of the theoretical treatises on unsteady gas flow, the solution for that behaviour in a pipe of varying area is found by solving the thermodynamic equations of state, continuity and energy together with the momentum equation[4,9,10]. The mathematics of these presentations are beyond reproach, except that they do not provide realistic solutions of either steeply tapered pipes or of flow regimes in converging sections.

The principal problem in this regard for tapered ducts is best seen by examining the previous paper, and in particular Section 2.3.7[1]. Here, as in many such treatises, the solution for change of wave pressure amplitude ratio, dX, is given by the classic statement caused by the 'correct'

solution of the thermodynamic equations of state, continuity and energy together with the momentum equation. In that equation the significant and controlling term is area change, i.e.

$$dX = function[dF/F] \qquad (1.1.1)$$

It is well known that, if the section taper of a diffuser is greater than 0.1 mm/mm, or an included angle of cone greater than about 6°, that exhaust diffusers in unsteady gas flow become less efficient. In reality the particle flow becomes two-dimensional as, in simpler words, flow separation from the wall takes place. In these circumstances the classical theoretical approach breaks down, as it very rapidly predicts sonic particle velocity and some form of theoretical corrective action is required or the computer program will crash arithmetically. That

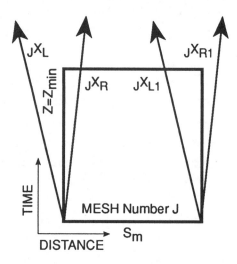

Fig.1 Solution of mass and state for each mesh

corrective action has already been proposed many years ago by Blair[11], by considering all such behaviour as a weak shock and then applying the Rankine-Hugoniot criteria to it to restore subsonic particle flow. For pipe tapers up to about 0.125 mm/mm, this solves the problem reasonably well. For pipe tapers greater than this value, the level of pressure wave reflection already predicted before the sonic velocity point is reached, and the corrective action taken, is so grossly in error that the calculation accuracy is always seriously impaired.

What is required is to solve the thermodynamic equations of state, continuity and energy together with the appropriate formulation of the momentum equation for unsteady flow within a duct which is either tapered, suddenly expanding, or of constant area, as a function not only of flow direction, but also its local potential for losses, i.e entropy gains, in terms of friction, heat transfer, and, more dramatically, flow separation

1.2 The meshing of the ducts. It is proposed to change the meshing of the ducts from that previously presented[1]. The change is shown in Fig.1, where each mesh segment is labelled and becomes virtually a theoretical 'island' of information. There are leftward and rightward pressure waves at each end of the mesh J in the duct at any instant in time. The mesh segment of volume V_J contains a gas of known properties (G_J and R_J) and purity PUR_J, at a pressure P_J, temperature T_J, density D_J, and mass M_J. The pressure waves calculated at either end cause mass and energy flow into or out of the mesh segment, in exactly the same manner as previously reported [1].

The new method of nomenclature for the computed pressure waves at each end of a mesh in a duct would seem to be a strategy which stores unnecessarily too much information. At the end of any time step, for the mesh J from the pressure wave propagation calculation, all that is known are the values of the rightward pressure wave at the righthand end of the mesh, $_JX_{R1}$, and the leftward pressure wave at the lefthand end of the mesh, $_JX_L$.

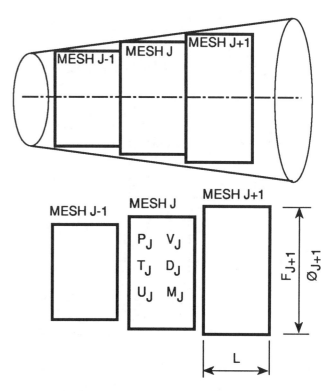

Fig.2 The meshing of the tapered duct

Simplistically, in a duct of parallel cross-section, if adjacent meshes J-1 and J+1 are at the same temperature as mesh J, i.e. if ,

$$T_{J-1} = T_J = T_{J+1}$$

then,

$$_JX_R = _{J-1}X_{R1}$$

$$_JX_{L1} = _{J+1}X_L$$

However, if the temperatures in the adjacent meshes are not identical, and that would be the normal situation, then the temperature discontinuity routine is applied, as clearly described by Blair[1] in Sect.2.3.8 of that paper.

1.3 Wave reflection and transmission in varying area ducts. Here the situation of particle flow for an expansion or a contraction in area must be analysed. The tapered pipe is meshed lengthwise as shown in Fig.2, and each mesh volume will be considered to be a duct of common diameter, with the adjacent meshes being naturally of a differing

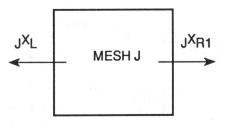

(a) knowns after "pipe" calculation

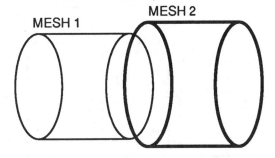

(b) adjacent pipe meshes for "unknowns" solution

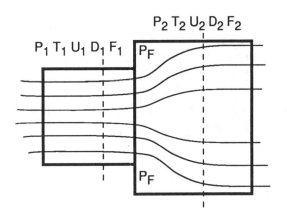

(c) the fluid-thermo situation

Fig.3 Gas flow from mesh to mesh.

diameter. As in Sect.1.2 above, at the end of any time step, for the mesh J from the pressure wave propagation calculation, all that is known are the values of the rightward pressure wave at the righthand end of the mesh, $_JX_{R1}$, and the leftward pressure wave at the lefthand end of the mesh, $_JX_L$. This is illustrated in Fig.3a. Particle flow in the rightward direction is considered by convention to be 'positive'.

In Fig.3b, adjacent meshes entitled 1 and 2 are illustrated, and the flow regime between them is shown in Fig.3c. Irrespective of the visual nature of that sketch, it is not stipulated whether the flow is contracting or expanding, simply that there will

be a face (or wall) pressure P_F assumed on the intermediate annular face for the flow to (or from) the gas at pressures P_1 and P_2. The traditional assumption (4-7) is that the face pressure for an expansion is the pressure at the minimum section, which would be the P_1 value in terms of rightwards flow in Fig.3c if the area F_2 is greater than F_1. That same assumption is then employed in those traditional solutions for contracting flow, as there is no possible provision in the mathematical solution for variation of the face pressure P_F. It may well be that this assumption is perfectly valid for both contracting and expanding flow regimes, however the solution proposed here is that the face pressure for expanding and contracting flows can be adjusted between the two extremes of P_1 and P_2 and be solved accurately at every section of the pipe system. The correct value of the face pressure for expanding and contracting flow during these unsteady flow regimes is currently under experimental investigation at QUB and this will be reported when it is complete; however, in practice in unsteady flow it is most unlikely that these face pressures will have equality for both flow directions. Equally, should flow separation be considered to occur, and this certainly can happen in diffusers, it is possible to introduce an alternative and high entropy gain assumption, i.e. by using a constant pressure relationship for the flow from 1 to 2, rather than the momentum equation; this is discussed further below.

It might be suggested that this analytic approach is tantamount to empiricism, but there is nothing empirical in solving precisely the thermodynamic and gas dynamic relationships at every section of flow in a duct. In that context, computational fluid mechanics is empirical in that it does just that in a finite element environment. Indeed, very the opposite is true, for in the traditional approaches (4-7), the single blanket assumption regarding the wall pressure in the solution of the momentum equation is empiricism, however complex and purist the ensuing mathematics may appear to be. It might also be queried that the proposed methodology is intensive in terms of computer cpu time. This is a less compelling argument today than hitherto, for the glib response is that accuracy is much more important than cpu time; the precise response is that the new calculation methodology given here is about 30% slower than that given previously, but the speed of the latest desktop PC or Macintosh™ is such that complete and complex engine design or analysis calculations, such as that recently reported to SAE (2), can still be accomplished in a few hours.

The equations of momentum, energy and continuity are solved between positions 1 and 2. The known parameters are $_1X_{R1}$, and $_2X_L$, and the unknowns are $_1X_{L1}$, and $_2X_R$.

The superposition pressures, temperatures, densities, acoustic velocities and particle velocities at positions 1 and 2 are given below. Note that superposition pressure is basically the summation of the leftwards and rightwards wave pressures, whereas particle velocity is a function of the difference between them.

Pressure amplitude ratio (superposition)-

$$_1X_{S1} = {_1X_{R1}} + {_1X_{L1}} - 1$$
$$_2X_S = {_2X_R} + {_2X_L} - 1$$

Pressure (superposition)-
$$P_1 = P_0 * (_1X_{S1})^{G7}$$
$$P_2 = P_0 * (_2X_S)^{G7}$$

Pressure (leftwards wave)-
$$P_{L1} = P_0 * (_1X_{L1})^{G7}$$
$$P_{L2} = P_0 * (_2X_L)^{G7}$$

Pressure (rightwards wave)-
$$P_{R1} = P_0 * (_1X_{R1})^{G7}$$
$$P_{R2} = P_0 * (_2X_R)^{G7}$$

Temperature-
$$T_1 = T_{01} * (_1X_{S1})^2$$
$$T_2 = T_{02} * (_2X_S)^2$$

Density-
$$D_1 = D_{01} * (_1X_{S1})^{G5}$$
$$D_2 = D_{02} * (_2X_S)^{G5}$$

Acoustic velocity-
$$A_1 = A_{01} * (_1X_{S1})$$
$$A_2 = A_{02} * (_2X_S)$$

Particle velocity-
$$U_1 = G5_1 * A_{01} * (_1X_{R1} - {_1X_{L1}})$$
$$U_2 = G5_2 * A_{02} * (_2X_R - {_2X_L})$$

It should be noted that the reference densities and acoustic velocities, D_{01} and D_{02}, and A_{01} and A_{02}, are given by,

$$D_{01} = P_0 / (R_1 * T_{01})$$
$$D_{02} = P_0 / (R_2 * T_{02})$$
$$A_{01} = \sqrt{(\gamma_1 * R_1 * T_{01})}$$
$$A_{02} = \sqrt{(\gamma_2 * R_2 * T_{02})}$$

and that T_{01} is not the same as T_{02} if an entropy gain is involved in the flow regime. Furthermore, the values of gas constant R and specific heat ratio γ and its functions G5 and G7 will also differ if the purity is dissimilar in meshes 1 and 2. However, in the small time steps which are used in the calculation, typically 60 μs, the purity from mesh to mesh, even in the exhaust system of a two-stroke engine, will not change so dramatically as to cause a significant error by using the properties of either mesh of an adjacent pair being analysed. However, it is possible within this calculation technique to select the correct values of gas properties for the flow regime, i.e. for particle flow from mesh 1 to 2, then the flowing gas has, and is given, the properties of the fluid in mesh 1; if the flow is reversed then the gas properties at mesh 2 are selected.

It can be seen that all of the above equations for the state conditions, or other parameters of velocity, contain the two unknown pressure amplitude ratio values, namely $_1X_{L1}$, and $_2X_R$. It is for these unknown values that the following expressions of continuity, energy, and momentum are solved.

Continuity equation from 1 to 2:
$$\Delta m = dt * D_1 * F_1 * U_1 = dt * D_2 * F_2 * U_2$$
$$(1.3.1)$$

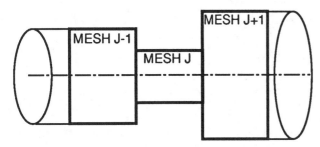

Fig.4 A throttle as sudden area changes

Energy equation for adiabatic flow from 1 to 2:

$$G5_1 * A_{01}{}^2 * {}_1X_{S1}{}^2 + U_1{}^2$$
$$= G5_2 * A_{02}{}^2 * {}_2X_s{}^2 + U_2{}^2 \qquad (1.3.2)$$

Momentum equation for flow from 1 to 2:

$$P_1 * F_1 + P_F * (F_2 - F_1) - P_2 * F_2$$
$$= (\Delta m/dt) * (U_2 - U_1) \qquad (1.3.3)$$

The value of P_F is a matter of debate, or perhaps experimentation is a better phrase, and the following is the current theoretical position for unsteady gas flow.

For contracting flow- $P_F = (P_1 + P_2)/2$

For expanding flow- $P_F = P_1$

The variable, either A_{01} or A_{02}, is eliminated from these equations, depending on the direction of flow, so that the gain of entropy is computed ultimately. One of these is known, depending on the flow direction; i.e. if the flow is from 1 to 2 then A_{01} is known, and *vice versa*. The resulting equation then contains only the unknowns of ${}_1X_{L1}$ and ${}_2X_R$.

The equations 1.3.1, 1.3.2 and 1.3.3 now contain the two unknowns of ${}_1X_{L1}$, and ${}_2X_R$ and can be solved by the Newton-Raphson method for two variables.

However, there are limits of applicability of these equations in expanding flow. For example, if the particle flow reaches a high Mach Number and the area ratio is either sufficiently large as to imply that it is a sudden expansion, or if the pipe taper is too steep, then flow separation from the walls of the diffuser may well occur. The limit criteria for these effects are discussed below in

Sect.3.1. In which case, as already found to be an effective solution by Benson(10), and also applied by Blair(1) for the cylinder-exhaust subsonic outflow boundary conditions, a constant pressure assumption should be used to replace the momentum equation, i.e.

$$P_1 = P_2 \qquad (1.3.4)$$

The equations 1.3.1, 1.3.2, and either 1.3.3 or 1.3.4, now contain the two unknowns of ${}_1X_{L1}$, and ${}_2X_R$ and can be solved in a Newton-Raphson solution for two variables.

These equations are formulated and solved anywhere in the ducting system wherever adjacent meshes are of dissimilar area. Perhaps the most significant theoretical contribution to computational accuracy is the mesh by mesh imposition of mass flow continuity and the indexing of the appropriate statement of the momentum equation for the actual local flow direction, or the indexing of an higher entropy gain (constant pressure) assumption should that be considered necessary at a given location. One of the most effective uses of the new approach to varying area flow, and in particular the concept of individual mesh by mesh computation of it, is that the calculation can handle the sudden change of area through a single mesh. An example of this would be a throttle or orifice buried within a pipe; this could be a throttle in an intake tract or the nozzle ring of a turbocharger. The potential geometry is shown sketched in Fig.4, and will be discussed later.

2. COMPUTATION EXAMPLES

Illustration of the use of the theoretical propositions above are given here. The Fig.5 shows pipes and throttles which will be employed theoretically for this purpose. The cylinder is considered to be filled with exhaust gas at 500ºC and at 1.6 atm pressure, and to be maintained at these conditions. The reference pressure P_0 is 1 atm and the atmospheric temperature beyond the pipe is at 20ºC. The pipe is filled initially with exhaust gas at 1 atm and 200ºC; i.e. the initial reference temperature in all of the duct meshes is

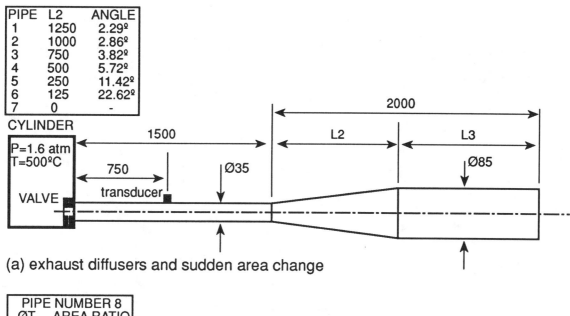

PIPE	L2	ANGLE
1	1250	2.29º
2	1000	2.86º
3	750	3.82º
4	500	5.72º
5	250	11.42º
6	125	22.62º
7	0	-

(a) exhaust diffusers and sudden area change

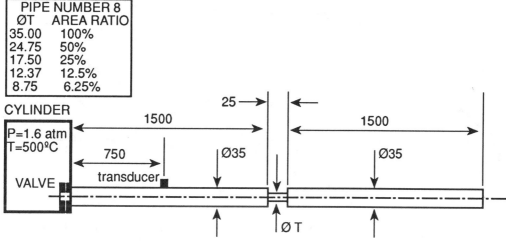

PIPE NUMBER 8

ØT	AREA RATIO
35.00	100%
24.75	50%
17.50	25%
12.37	12.5%
8.75	6.25%

(b) throttles within a pipe ducting

Fig.5 The several pipes, diffusers and throttles for the simulations.

T_0 which is at 200ºC. The pipe wall is also at 200ºC. The cylinder is of infinite volume, i.e. upon outflow the pressure stays constant at 1.6 atm, and the exhaust port area increases linearly with time to equal 80% of the exhaust pipe area by 2.5 ms; it shuts in the same fashion in 2.5 ms. The first section of the exhaust pipe in all cases is 35 mm diameter and is 1500 mm long. The pipe is selected at this length so that the pressure waves are well spaced in time and distance and are easily recognised with regards to their formation, or by their reflection from any further and onward changes of section.

2.1 The diffuser pipes. The diffuser pipes are described in Fig.5a. The total length of the final section of pipe is L2 plus L3, and is fixed at 2000 mm long and always finishes at 85 mm diameter. The length of the diffusers L2 employed for pipes 1-6 are 1250, 1000, 750, 500, 250, and 125 respectively. This gives tapers for the cones in pipes 1-6 of 0.04, 0.05, 0.075, 0.1, 0.2 and 0.4 mm/mm respectively. The included angles of the diffusers in pipes 1-6 are 2.29º, 2.86º, 3.82º, 5.72º, 11.42º, and 22.62º respectively. Traditionally, pipe number 4 would be regarded as the widest taper diffuser which in practice would function efficiently as an exhaust diffuser in unsteady gas flow. This would be particularly true for four-stroke cycle engines where the exhaust pulse is longer in time, and often of greater amplitude, than that in a two-stroke

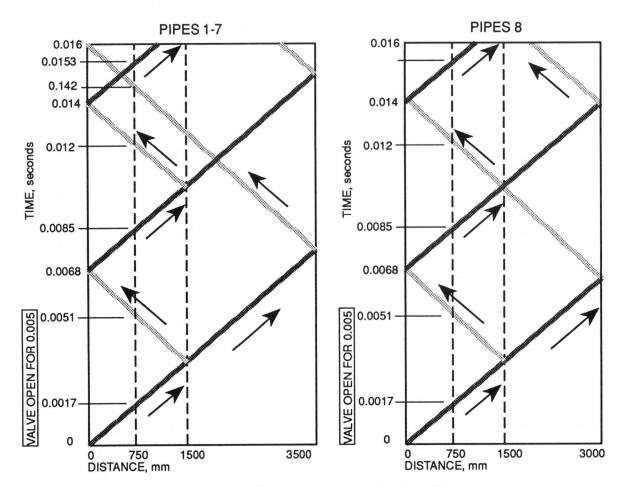

Fig.6   Time-distance characteristics for acoustic waves in pipes 1-8

engine.

The shorter diffusers in pipes 5 and 6 would begin to approximate in practice to sudden expansions from the 35 mm down pipe diameter to the 85 mm tail pipe diameter.

2.2   The sudden expansion pipe.   The pipe numbered 7 in Fig.5a has a sudden expansion at the end of the 1500 mm long down pipe, i.e. L2 is zero and L3 is 2000 mm.

2.3   The buried throttles.   In Fig.5b is shown a series of pipes all numbered 8, but which have a one mesh long, i.e. 25 mm, throttle at 1500 mm from the 'cylinder' and 1500 mm from the open end to the atmosphere.   The throttle area ratios range from 100%, i.e. no throttle at all and in effect giving a straight pipe of 3025 mm in length, to 50%, 25%, 12.5% and 6.25% area ratio. The corresponding pipe mesh diameters ØT at the

throttle section are 35 mm, 24.75 mm, 17.5 mm, 12.37 mm and 8.75 mm respectively.

2.4   The recording position.   For all of the above geometries of pipes 1-8, at 750 mm in the middle of the down pipe, there is assumed to be a recording position for pressure and particle velocity.   The signals noted for all of the computations are superposition pressure, leftwards and rightwards wave pressures, and the particle velocity.   In experimental practice, only the superposition pressure could be recorded with any degree of surety by employing a fast response pressure transducer.   In short, the computation provides information which would be difficult, or even impossible, to determine experimentally with reasonable accuracy.   To aid the reader in understanding the possible wave transmission and reflection behaviour in the two pipe systems, a simple acoustic distance-time characteristic is

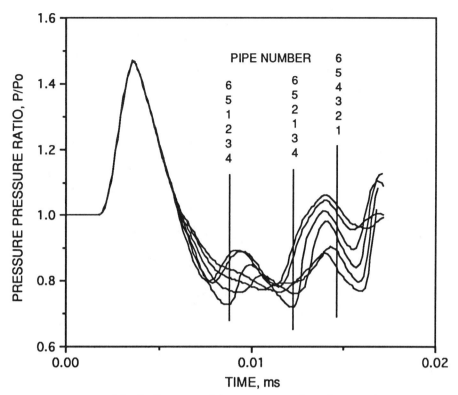

Fig.7 Superposition pressure for pipes 1-6

shown illustrated in Fig.6.

2.5 The presentation of the computed results. At the recording position there is plotted the pressure-time and velocity time characteristics for all of the pipe systems 1-8 discussed above. Time is stopped at about 17 ms, when most of the dynamic effects have begun to be attenuated by friction, or to wane by transmission outward to the atmosphere.

In Figs.7-10 are the superposition pressure, rightwards wave pressure, leftwards wave pressure and particle velocity characteristics as a function of time for the diffuser pipes 1-6 respectively. The gas particle velocity is shown as Mach number. The dashed line with the annotated pipe numbers indexes the P-t or M-t curves and the vertical order of precedence denotes the pipe number associated with that particular characteristic.

In Figs.11-14 are the superposition pressure, rightwards wave pressure, leftwards wave pressure and particle velocity characteristics as a function of time for the sudden expansion pipes 6 and 7.

In Figs.15-18 are the superposition pressure, rightwards wave pressure, leftwards wave pressure and particle velocity characteristics as a function of time for the throttled cases described as pipes 8. The dashed line with the annotated throttle area ratios (as percentages) indexes the P-t or M-t curves and the vertical order of precedence denotes the throttle area ratio associated with the particular characteristic.

3. DISCUSSION OF RESULTS

3.1 For the diffuser systems shown in Figs.7-10. As can be seen in Figs.7 and 8, the exhaust pulse is some 1.5 atm in amplitude and prevails for the 5 ms when the valve is opened. It arrives at the 'transducer' at 750 mm down the pipe some 1.7 ms after the valve opening point when time is zero. These effects can be seen in simpler terms on the characteristic diagram in Fig.6. Some steepening of the wave front can also be observed. The first point of the reflection from the diffuser can be seen to arrive at the 'transducer' by about 6 ms.

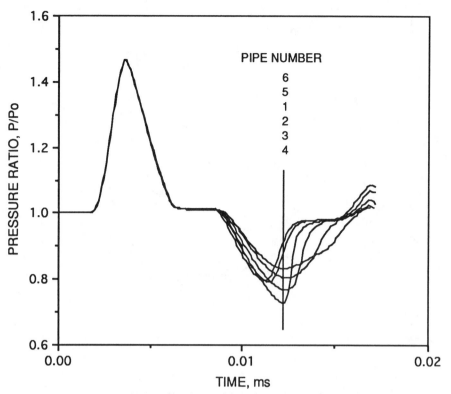

Fig.8 Rightwards pressure waves for pipes 1-6

The best view of the reflection of the exhaust pulse from the diffusers can be found in Fig.9, by examining the leftwards wave pressures. Pipe 1, with the longest diffuser, has the longest in time terms, but the shallowest in pressure terms, sub-atmospheric reflection. In Fig.10 it can be seen to extract exhaust gas further down the pipe, in the same direction as the main exhaust pulse. As the diffusers become shorter and the tapers are steeper, the sub-atmospheric reflection deepens and the associated particle Mach number rises. The deepest reflection is given by pipe 4, with its 0.1 mm/mm taper and a 5.72º included angle of cone.

The next steeper diffuser, pipe 5, can be seen to be less effective in this regard. This trend continues with pipe 6, which has just 125 mm of diffuser length. Wall separation has been deemed to occur and the calculation has indexed from time to time, and from place to place, the constant pressure equation instead of the momentum equation.

If these calculations had been conducted by the older Riemann variable methods, as previously

shown by Blair(11) for diffuser systems, the dF/F term employed within the prediction for the dλ and dβ terms in the steep diffusers in pipes 5 and 6, would have given exaggerated values of suction reflection and particle Mach number. As has already been mentioned in Sect.1.1, in the traditional calculations using the method of characteristics the pipe 5 would have 'achieved' a particle velocity of Mach 1 somewhere in the first few calculation meshes of the diffuser and a 'shock criteria' employed (11) as an arithmetic and thermodynamic hammer to eliminate it. In the new approach, the calculation is monitoring both the wall angle of the duct with respect to distance and the local particle Mach number; should the local Mach number exceed a limit criteria, typically 0.7, and the wall angle be greater than its limit criteria, typically 7º included, then the thermodynamic analysis at that particular mesh boundary in the duct will select the 'constant pressure equation', Eqn.1.3.4, to be included in the mathematical set rather than the momentum equation, Eqn.1.3.3. With this new theoretical approach, not only is the solution non-

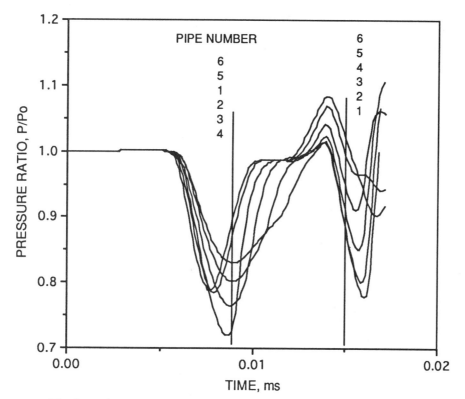

Fig.9 Leftwards pressure waves for pipes 1-6

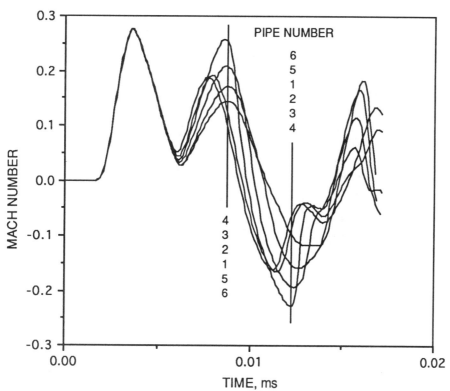

Fig.10 Gas particle Mach Number for pipes 1-6

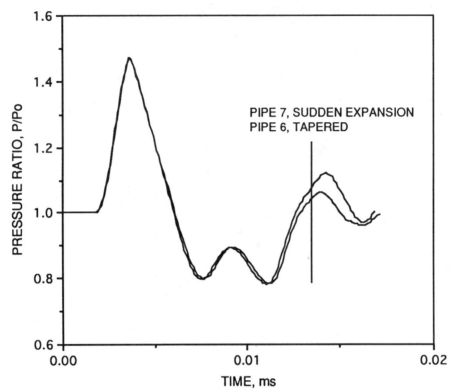

Fig.11 Superposition pressure waves for pipes 6 and 7

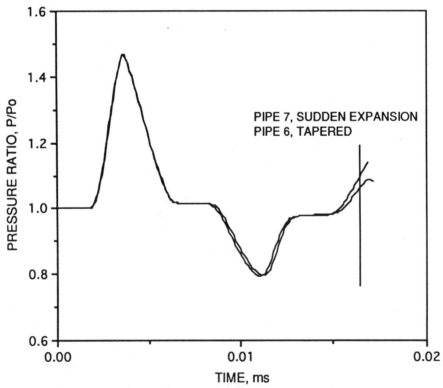

Fig.12 Rightwards pressure waves for pipes 6 and 7

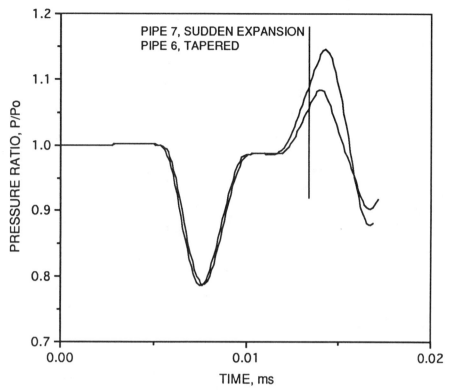

Fig.13 Leftwards pressure waves for pipes 6 and 7

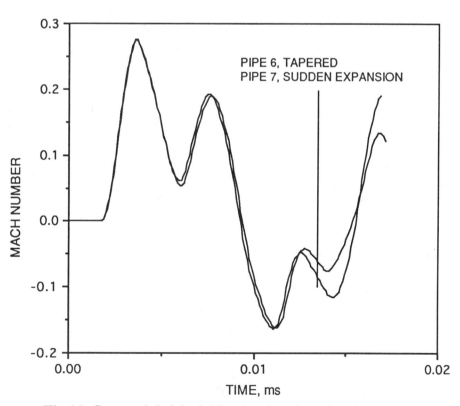

Fig.14 Gas particle Mach Number for pipes 6 and 7

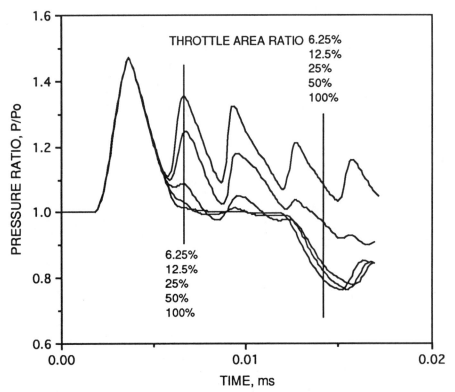

Fig.15 Superposition pressure waves for throttled pipes 8

isentropic, but the appropriate momentum equation is indexed with regards to the flow diffusing or contracting, or the constant pressure criteria is employed should the area ratio and the particle velocity from mesh to mesh become excessive. The result is a realistic prediction of the flow in diffusers and nozzles, however steep they may be.

3.2 Comparison of the steepest diffuser and a sudden expansion in Figs.11-14. The shorter the diffuser becomes, the more it should behave as a sudden expansion. In the Figs.11-14, it can be seen that pipe 6, the shortest diffuser, and pipe 7, the sudden expansion, produce virtually identical P-t and M-t diagrams. Only at the second reflection, of the suction wave bouncing off the closed end and arriving again at each of these section changes in their respective pipes, does their behaviour begin to differ somewhat.

3.3 Comparison of the various throttle area ratios in a parallel pipe in Figs.15-18. The effect of throttling in a parallel pipe can be clearly observed. As the throttle area ratio becomes smaller, then the leftward wave pressures show increasingly larger compression waves as reflections off the obstruction in the pipe. Naturally, the 100% throttle area ratio means that there is no throttle there at all, and effectively that particular version of pipe 8 is a straight pipe of 3025 mm length. The 50% throttling gives compression wave reflections but the throttled section is reasonably transparent to the suction reflection from the final end of the pipe. That is still true for the 25% throttle area ratio. However, at 12.5% a major shift takes place and the throttle gives a much greater choking effect for all reflections, sending mirror image pressure signals each time any pressure wave encounters the considerable throttle restriction. This effect is further exaggerated at the 6.25% throttle area ratio and the rising mean pressure level is caused by choked flow through the severe restriction of the 8.75 mm diameter within the 35 mm diameter pipe.

It is very interesting to note the major shift in behaviour between the 25% and the 12.5% nozzle

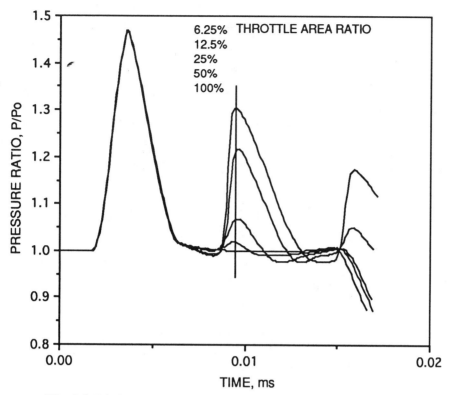

Fig.16 Rightwards pressure waves for throttled pipes 8

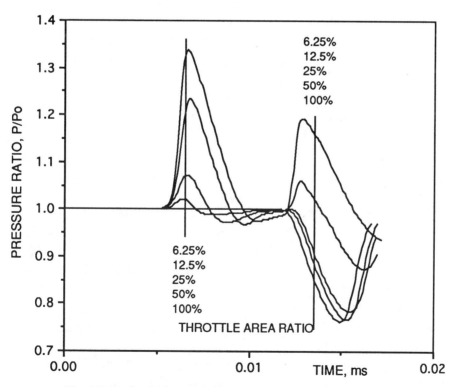

Fig.17 Leftwards pressure waves for throttled pipes 8

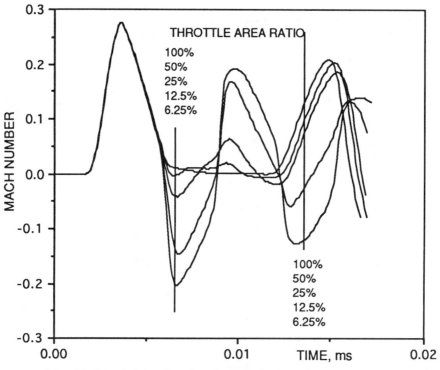

Fig.18 Mach Number for throttled pipes 8

area ratio, particularly as this form of restriction within an exhaust pipe is geometrically analogous to that posed by the nozzle ring of a turbocharger. The successful and traditional design criterion for the area ratio of the turbocharger nozzle ring is between 20 and 25%, and anything less has been shown to reduce the turbine work available. The ensuing outcome is diminished energy delivered into the compressor(12). It is interesting to note that there is some unsteady gas dynamic logic behind that empiricism.

Realistic modelling of the complete turbocharger, i.e both the turbine and the compressor, which is located within the overall ducting system of a simulated engine, is now a distinct possibility by virtue of this method of analysis.

Clearly, this form of theoretically modelling a restriction also applies to intake throttles in induction systems, and indeed that would be its most common utilisation. Intake throttle modelling has been carried out previously, but not quite so realistically as it is presented here(13,14).

4. CONCLUSIONS

A non-isentropic model of unsteady gas flow in varying area ducts, in sudden expansions or contractions in ducts, and in orifices within ducts, has been produced and its utilisation has been successfully demonstrated.

The theory has been clearly described, and the vital importance is emphasised of, with respect to calculation accuracy, not only imposing mass continuity and energy conservation at each mesh junction, but also selecting the correct formulation of the momentum equation for the particular flow regime in question.

For situations where entropy gains are inevitable, such as expansion flow in steeply tapered pipes where separation is occurring, or at sudden expansions in a pipe, then the use of a constant pressure criterion is justified; it must be emphasised in that event that mass flow continuity and energy conservation are still required to be imposed.

In this publication it has been shown to predict logically, and with arithmetic stability, the behaviour in steeply tapered ducts which hitherto, as far as the author is aware, has not been possible with other calculation techniques. In this context, it should be recalled that this calculation method has already been used to predict, very accurately, the unsteady gas flow in, and overall performance characteristics of, both two-stroke and four-stroke cycle engines(2,8). It is important that experimental work is conducted to verify many of the theoretical contentions made in this paper, and in those previous publications, relating to this computational model of unsteady gas flow proposed by Blair(1). This work is now under way at QUB.

The theory presented here, together with that given previously (1), now makes this calculation technique quite uniquely different from any previously proposed in the literature.

5. ACKNOWLEDGEMENTS

The authors would like to acknowledge The Queen's University of Belfast for providing both the theoretical environment and the accompanying experimental and manufacturing facilities without which a real learning process in engineering is not possible. There are bean-counters in all governments, dedicated to pursuing the 'efficient utilisation of resources' who believe, doubtless with the sincerity born of ignorance, that it is possible; they are wrong.

6. NOTATION

NAME	SYMBOL	(units)
Area	F	(m²)
Density	D	(kg/m³)
Diameter	Ø	(m)
Gas constant	R	(J/kgK)
Length	L	(m)
Mach Number	M	
Mass	m	(kg)
Mass flow rate	M	(kg/s)
Pressure	P	(Pa,atm)
Pressure Amplitude Ratio	X	
Pressure Ratio	PR	
Purity	Π	
Ratio of Specific Heats	γ	
Temperature	T	(K)
Time	t	(s)
Velocity (particle)	U	(m/s)
Velocity (acoustic)	A	(m/s)
Volume	V	(m³)

Subscripts-	
superposition	S
isentropic reference	0
leftward	L and L1
rightward	R and R1
face or wall	F
mesh positions	J, 1 and 2

Gamma Function Number-	
function 5	$G5 = 2 / (\gamma - 1)$
function 7	$G7 = 2 * \gamma / (\gamma - 1)$

7. REFERENCES

1. G.P.Blair, "An Alternative Method for the Prediction of Unsteady Gas Flow through the Reciprocating Internal Combustion Engine", Society of Automotive Engineers International Off-Highway & Powerplant Congress, Milwaukee, Wisconsin, September 9-12, 1991, SAE Paper No.911850 and in SP883, pp24.
2. G.P. Blair, "Correlation of an Alternative Method for the Prediction of Engine Performance Characteristics with Measured Data", Society of Automotive Engineers, International Congress, Detroit, Michigan, March 1993, SAE paper

No.930501 and in SP945, pp20.

3. G.P.Blair, J.R.Goulburn, "The Pressure-Time History in the Exhaust System of a High-Speed Reciprocating Internal Combustion Engine", Society of Automotive Engineers, Mid-Year Meeting, Chicago, Illinois, May 15-19,1967, SAE Paper No.670477.

4. R.S.Benson, R.D.Garg, D.Woollatt, "A Numerical Solution of Unsteady Flow Problems", Int.J.Mech.Sci., Vol.1, p253, 1960

5. M.Poloni, D.E. Winterbone, J.R. Nichols, "Comparison of Unsteady Flow Calculations in a Pipe by the Method of Characteristics and the Two-Step Lax-Wendroff Methods", Int.J.Mech.Sci., Vol.29, 1987.

6. J.D. Ledger, "A Finite Difference Approach for solving the Gas Dynamics in an Engine Exhaust", Int.J.Mech.Sci., Vol.17, No.5, p271, 1975.

7. M.Chapman, J.M.Novak, R.A.Stein, "A Non-Linear Acoustic Model of Inlet and Exhaust Flow in Multi-Cylinder Internal Combustion Engines, Winter Annual Meeting, ASME, Boston, Mass., 83-WA/DSC-14, November 1983.

8. G.P. Blair, "Correlation of Measured and Calculated Performance Characteristics of Motorcycle Engines", Funfe Zweiradtagung, Technische Universität, Graz, Austria, 22-23 April 1993.

9. G. P. Blair, "The Basic Design of Two-Stroke Engines", Society of Automotive Engineers, Warrendale, Pennsylvania, February 1990, pp672, SAE ref. no. R-104, ISBN 1-56091-008-9.

10. R.S.Benson, "The Thermodynamics and Gas Dynamics of Internal Combustion Engines", Volumes 1 and 2, Clarendon Press, Oxford, 1982.

11. G.P.Blair, M.B.Johnston, "Unsteady Flow Effects in the Exhaust Systems of Naturally Aspirated, Crankcase Compression, Two-Stroke Cycle Engines", SAE Farm, Construction and Industrial Machinery Meeting, Milwaukee, Wisconsin, September 11-14,1968, SAE Paper No.680594.

12. G.P. Blair, "Unsteady Flow Characteristics of Inward Radial Flow Turbines", Doctoral thesis, The Queen's University of Belfast, May 1962.

13. J.F.Bingham, G.P.Blair, "An Improved Branched Pipe Model for Multi-Cylinder Automotive Engine Calculations", Proc.I.Mech.E., Vol.199, No.D1,1985, pp65-77.

14. A.J.Blair, G.P.Blair, "Gas Flow Modelling of Valves and Manifolds in Car Engines", Proc.I.Mech.E., International Conference on Computers in Engine Technology, University of Cambridge, April 1987, C11/87, pp131-144.